Tourism and Memory

I0131796

This book considers tourism to memorial sites from a visitor's point of view, challenging established theories in tourism and memory studies by critically appraising Germany's often celebrated memory culture.

Based on visitor observations and exit interviews, this book examines how domestic and international visitors negotiate their visits to the concentration camp memorials Ravensbrück and Flossenbürg, the House of the Wannsee Conference and the former Stasi prison Bautzen II. It argues that memorial sites are melting pots where family, national and global narratives meet. For German visitors, the visit to memorial sites is a confrontation with Germany's responsibility for the two dictatorships, while for international visitors it can be a form of 'seeing is believing'. Ultimately, the immediacy of the memorial sites is required to instil democratic values in future generations.

Rooted in an interdisciplinary approach, this book will be of interest to academics and students in German studies, tourism and heritage studies, museum studies, public history and memory studies.

Doreen Pastor has completed her PhD in German studies at the University of Bristol and is currently a lecturer in the Department of Politics, Languages and International Studies at the University of Bath. Her research focuses on the memorialisation of the Nazi and GDR past in contemporary Germany, particularly how visitors engage with a challenging past in museums.

Routledge Studies in Second World War History

The Second World War remains today the most seismic political event of the past hundred years, an unimaginable upheaval that impacted upon every country on earth and is fully ingrained in the consciousness of the world's citizens. Traditional narratives of the conflict are entrenched to such a degree that new research takes on an ever important role in helping us make sense of World War II. Aiming to bring to light the results of new archival research and exploring notions of memory, propaganda, genocide, empire and culture, Routledge Studies in Second World War History sheds new light on the events and legacy of global war.

Recent titles in this series

War Through Italian Eyes
Fighting for Mussolini, 1940–1943
Alexander Henry

The Novel *Das Boot*, Political Responsibility, and Germany's Nazi Past
Dean J. Guarnaschelli

World War II Historical Reenactment in Poland
The Practice of Authenticity
Kamila Baraniecka-Olszewska

Tourism and Memory
Visitor Experiences of the Nazi and GDR Past
Doreen Pastor

Shoah and Torah
David Patterson

For more information about this series, please visit: www.routledge.com/ Routledge-Studies-in-Second-World-War-History/book-series/WWII

Tourism and Memory

Visitor Experiences of the Nazi and GDR Past

Doreen Pastor

Routledge
Taylor & Francis Group

LONDON AND NEW YORK

First published 2022
by Routledge
2 Park Square, Milton Park, Abingdon, Oxon OX14 4RN

and by Routledge
605 Third Avenue, New York, NY 10158

Routledge is an imprint of the Taylor & Francis Group, an informa business

© 2022 Doreen Pastor

British Library Cataloguing-in-Publication Data
A catalogue record for this book is available from the British Library

Library of Congress Cataloging-in-Publication Data
A catalog record for this book has been requested

ISBN: 978-0-367-64895-4 (hbk)
ISBN: 978-1-032-00499-0 (pbk)
ISBN: 978-1-003-12683-6 (ebk)

DOI: 10.4324/9781003126836

Typeset in Bembo
by Apex CoVantage, LLC

In memory Gertrud Schnabel and Hannelore Pastor

Contents

Acknowledgements

Since the book is based on my Ph.D. research, I would like to thank my supervisors, Dr Debbie Pinfold and Dr Mark Allinson at the University of Bristol, for their patience and support, without which I would have given up when dealing with stumbling blocks.

A huge 'thank you' goes to the memorial managers who allowed me to conduct the visitor research:

- Dr Jörg Skriebeleit Flossenbürg Concentration Camp Memorial
- Dr Matthias Heyl/Dr Insa Eschebach Ravensbrück Concentration Camp Memorial
- Dr Elke Gryglewski House of the Wannsee Conference
- Silke Klewin/Dr Bert Pampel Bautzen II Stasi Prison Memorial

And of course, I would like to thank all the visitors who so patiently answered my questions and spent time with me. Two visitors sent an e-mail to Flossenbürg's management team, saying that they very much enjoyed talking to the 'young researcher'. They probably did not know how important this e-mail was. It kept me going during tough days.

I would also like to thank Nick and Val Drew, who supported me ever since I met them during my exchange year at Bath Spa University in 2005. I am also indebted to Professor Stephen Ward, who was always on-hand to proofread my work and whose enthusiasm for my research kept me going during difficult periods.

Last but not least, I would like to thank my Mum, Kerstin Pastor. She supported the fieldwork in Germany logistically by transporting suitcases, picking me up from stations and supplying me with food. I still remember the day she arrived with coffee and cake in Flossenbürg. Thank you so much. I am so grateful, and I know that I do not always show it.

1   Tourism to memorial sites

'It is perverse: this morning I had a great breakfast at the hotel, now I'm looking at this [Ravensbrück] and then I carry on with my cycle ride.' This statement was made by a visitor at Ravensbrück during his walk across the memorial site. His visit was clearly embedded within a tourist itinerary. In fact, he used the new cycle path that links Berlin with Copenhagen. It is a cycle ride that leads him through the unspoilt landscapes of the north German lake district, known as the *Mecklenburger Seenplatte*, and whose tourism strategy promises *Endlich Ruhe* (finally peace). For the tourist who steps onto the grounds of Ravensbrück, the slogan must seem perverse. Within the space of a couple of minutes, s/he is confronted with the display of horrific atrocities.

For many academics (Lennon and Foley 2010; Pollock 2003; Cole 1999), and indeed not just academics, the concept of adding memorial site visits to an overall tourism itinerary is inappropriate. Griselda Pollock (*ibid*) interprets tourism as inherently voyeuristic, with Holocaust tourism taking it to a new and terrifying level. It is the suffering of others that is made into a spectacle. In a similar vein, Tim Cole (1999) states that tourists to Auschwitz visit a theme park where they have little awareness about what is (or is not) authentic. The fiercest criticism often comes from the survivors themselves, with Ruth Klüger (2001), for instance, arguing against a camp culture that merely fulfils our necrophiliac desires. Yet Pollock (2003) asserts that her son's educational visit to Auschwitz is justified since it was based on serious encounters with the site. Thus, the debate about tourism to memorial sites is not just one about (in) appropriate behaviour, it is also about who should (or should not) visit.

No other group suffers more persistent criticism than the tourist (Culler 1981). The origins of this negative perception of the tourist can be traced back to the rise of mass tourism, with authors (e.g. Fussell 1982) lamenting the decline of the 'traveller' and the increase of the superficial 'tourist'. Dean MacCannell (1973) argues that it is chic to intellectually condemn tourists. In the academic literature, this viewpoint leads to attempts to distance oneself from the tourist. When Cole (1999, 115), for instance, writes 'I like to think I went to Auschwitz with the loftier intentions of the "pilgrim"', he purposefully places himself into an intellectually superior position that allows him to criticise the 'ordinary' tourist.

DOI: 10.4324/9781003126836-1

Piotr Cywiński (2019), the memorial manager for Auschwitz-Birkenau, poignantly summarises the difficulties with such opinions:

> Everyone thinks that they are the most aware, that they know how to behave in the best and most sensitive way, what to ask, when to nod and keep silent. Other people are a distraction. When you are standing face to face with humanity, hell is indeed other people – especially those we suspect of being tourists or of having tourist intentions, tourist opinions, tourist needs.
>
> (*ibid*, 18)

When analysing tourism to memorial sites, the academic discourse is dominated by research into 'dark tourism' which focused on the transactional relationship between the 'dark' site and the consumer 'tourist'. John Lennon and Malcolm Foley (2010), who coined the term, explained that the rising visitor numbers are largely driven by the media; hence, only sites that are still in living memory can be regarded as 'dark tourism' sites. Philip Stone (2011) subsequently widened the definition by suggesting that all journeys to sites of death and the macabre are a form of dark tourism. His core claim is that Western societies' inability to deal with death in the public sphere leads to a fascination with visiting sites of atrocities. Such explanations are, however, too simplistic. Memorial sites do not only focus on the past or death; they can also be future-orientated and committed to truth and reconciliation. Both South Africa and Northern Ireland frame their past struggles, albeit in different ways, under the premise of a more hopeful future (Lisle 2016). By contrast, Light et al.'s (2020) research in Romania highlighted that memorial sites can be active participants in transitional justice processes. Moreover, Jennifer Hansen-Glucklich (2014) emphasises that Holocaust exhibitions are framed within the wider national narrative. Whilst Berlin's Holocaust exhibition at the Jewish Museum is focused on the emptiness the atrocities left behind, Yad Vashem in Israel and the Holocaust Museum Washington are framed within positive nationalist narratives. Yad Vashem stresses death and destruction but ends with the birth of Israel. At USHMM, however, the permanent exhibitions depict the Americans as liberators and the United States as a haven for democratic values. Therefore, 'difficult' exhibitions might deal with death, yet the messages they impart on visitors differ.

Thus, sites that tell the story of nation formation have considerable agency (Bruner 2005). They are not silent places and can generate meaning through the tourist's responses to the site. This meaning is influenced by political and social discourses that change over time. The first scholar who recognised that memory is not just an individual process was Maurice Halbwachs (1925) with his concept of 'mémoire collective'. For him, memories are socially framed, thus cementing the foundation of groups. Consequently, understanding an individual's thinking process requires an analysis of the group dynamics. Since Halbwachs' founding theory, a variety of terms have been used to describe the use of the past by social groups; yet there is currently no generally accepted definition of the term (Moller 2010). In cultural studies, Aleida Assmann (2011) influenced the academic discourse with the introduction of the terms 'cultural' and 'communicative' memory, with the latter referring to the living memories that are transmitted between generations and the former focusing on fixed memory such as museums.

James Fentress and Chris Wickham (2012) argue that for followers of Halbwachs' theory it is always difficult to locate the individual in memory processes without reducing him/her to a passive object entirely dependent on the collective. Similarly, Assmann (2011), with her focus on cultural memory, views collective memories as fixed and thus risks making the same mistake as Halbwachs (Robinson 2019). In so doing, research in memory studies tends to focus on the explicit traces of memory and neglects human interactions. As early as 1997, Alon Confino, therefore, suggested considering the ways people construct the past by including social practices and transmission, specifically mentioning tourism.

Anthropologists, by contrast, are naturally more concerned with the 'webs people construct' (Geertz 1973, 5) to uphold or form identities, and as such, their interest is focused on how memory is used during meaning-making processes. Since memories define our sense of self and our connection to the group, individual narratives are important dimensions for anthropological research (Cattell and Climo 2002). Moreover, Maria Cattell and Jacob Climo (*ibid*) point out that, in a highly mobile world, memories have to be adapted to fit new meaning-making systems. This link between social discourse and individual narrative was shown in Barbara Einhorn's (2000) work on life biographies in which German women adapted their memories several times in order to adjust to the Nazi regime, then to the GDR and finally to contemporary Germany. Geoffrey White's (1999) research project at Pearl Harbour and Eric Gable and Richard Handler's (2000) research at Colonial Williamsburg have shown that a museum visit is a careful negotiation between the family and the national narrative, which helps to cement a sense of self. Thus, memory is a fluid process that shifts according to the social and group contexts. Researching tourists at former sites of atrocities, therefore, requires the uncovering of the negotiation process between the self, the family narrative and the national discourse. The German author Navid Kermani (2018) illustrates this argument in his book *Entlang den Gräben: Eine Reise durch das östliche Europa bis nach Isfahan* (*Along the Trenches: A Journey through Eastern Europe to Isfahan*), in which he describes his arrival at the memorial site Auschwitz–Birkenau. Whilst most tour groups meet in front of the *Arbeit macht frei* gate for a group photo, his German group waited silently and ashamed for the tour guide away from the gate. His visit also involved the interrogation of his mixed Iranian and German heritage, leading to questions of shared responsibility for past German crimes in contemporary Germany. He finally concluded that a German visitor will never be just a casual visitor at Auschwitz.

Hence, visits to memorial sites are shaped by contemporary societal discourses. For example, Anne Hertzog (2019) explains that tourists travelling to the First World War battlefield sites in the 1920s would have had a very different experience from those travelling today, as memories of the conflict would have been raw and thus still shaped by national hostilities. Today, Caroline Winter's (2015) analysis of visitor comments in the guest books at Tyne Cot Cemetery in Belgium, the largest Commonwealth War Graves Commission Cemetery in the world, revealed a ritualised commemoration of the dead, shaped by the site itself and the cultural background of the visitor. Even though today's former battlefield sites share little resemblance with the muddy trenches during the First World War, the importance of these landscapes lies in their 'embodied memories'.

The concept of embodiment entails two aspects: embodied cognition and embodied emotions (Niedenthal and Maringer 2009). The latest research in neuroscience argues that behaviour is not only generated within the brain but also develops in real time as an interaction between the environment and the body. Edward Casey (1987) emphasised that bodies contain memories that are activated by encounters with meaningful places or symbols, which is particularly relevant to memorial sites. A research project at the Nanjing Massacre Memorial in China revealed that visitors responded to temperature changes or dim lighting, making them feel sad, cold or mournful (Xie and Sun 2018). Similarly, Shanti Sumartojo's (2019) qualitative research at Camp de Milles, a concentration camp memorial in France, has shown that the materiality of the space evokes personal memories, triggered by atmospheric and sensory experiences with the site. Thus, the symbolic meaning induced by the site influences visitors' behaviour. If a visitor trembles when walking into a crematorium, as evident in my research outlined in later chapters, then symbolic meaning, aesthetics, foreknowledge and personal memories form a bond that engenders an embodied reaction.

In fact, a visitor at a memorial site such as a concentration camp has to actively use his/her imagination as the sites are marked by absence (Popescu 2016). Imaginaries are ingrained in people's family backgrounds and in a vision of the world, of its people and of its places and are culturally shared (Gravari-Barbas and Graburn 2012). Tourism would, indeed, not take place without these imaginaries as they determine how a destination is chosen and, once onsite, which places to visit or avoid. Whilst Maria Gravari-Barbas and Nelson Graburn focus largely on the leisure industry, memorial sites are equally shaped by these imaginaries. Patrice Keats (2005, 183) explains that the gaps at concentration camp memorials (in this case Auschwitz-Birkenau) are filled with inner images, which makes the visit clearer, more vivid and more detailed; yet it is constrained by what the visitor knows. In addition, these tourism imaginaries are carefully constructed by local stakeholders. Hertzog (2012), for instance, notes that the Picardie's regional identity as the 'sacrificed and wounded land' has been reconstructed around the scarred landscape of the First World War and is promoted as such. Since place as a complex with physical, social, cultural and emotional features is a central part of the tourism experience (Rickly-Boyd 2012), it requires an analysis that uncovers these layers. After all, as Jeff Malpass (2011, 14) asserts, 'understanding landscapes means understanding the forms of actions out of which they arise', thus highlighting the performative aspect of the place. The concept of performativity gains importance in the humanities and social sciences in order to explain how our lived worlds and identities are constantly shaped and reshaped through our actions and gestures in a social space (Schult 2018).

MacCannell (1989) also reminds us that tourist attractions are not random assemblies of artefacts. Firstly, they have to be identified as worthy of preservation, then named and framed, and finally mechanically (e.g. with photographs) and socially reproduced before they appear on a tourist itinerary. Memorial sites were certainly set aside and nationally framed to invite people to grieve for lives lost. In addition, memorial sites are usually designed to create an effect. For example, the first memorial at Flossenbürg Concentration Camp was designed in the way of the cross with a descent into hell (the crematorium) and an ascent to salvation (the chapel), using the natural backdrop of a valley that was subsequently called the 'valley of death'. Hence, the landscape was deliberately manipulated to emphasise the suffering of the victims using Christian symbolisms. Judith Wasserman (1998) argues that memorials often use natural features that are steeped in ancient rituals, for example, water or plants, which turn the visitor into an active participant in memory processes. Ravensbrück is another case in point as trees that were planted during the camp's operation were retained as 'natural eyewitnesses' during the recent landscape design process (Tischer 2020, personal communication). Thus, Wasserman (*ibid*) suggests that memorials will not only deepen intellectual knowledge, they will also allow visitors to gain experiential insights.

Sumartojo (2016), by observing Anzac Day in Australia, goes further by emphasising that one cannot just analyse commemorative events by considering the historical, symbolic or textual contexts; one has to take the effects of atmosphere into account. Atmosphere is a complex interplay between the architecture of a memorial site, the symbolic value of the surrounding area, light and dark, the presence or absence of other people, the weather conditions, and so forth. The design of a memorial site can enhance an atmosphere – for example, by covering the former area of the barracks at Ravensbrück with a black surface made of clinker, the site creates the illusion of a 'shadowy' place. Sumartojo (*ibid*) also highlights that atmosphere is influenced by foreknowledge and anticipation: the more visitors knew about Anzac soldiers, the more affectively charged the commemorative event is. This sensory experience, as I will show throughout the book, is a crucial component of the visitor's experience. It is therefore vital to understand how tourists ascribe meaning to the site because 'without meaning, memory is nothing' (Hunt and McHale 2007, 42).

Tourism inherently also involves an encounter with the 'other' as tourists will engage with new environments, cultures, societies and peoples (Thomas 2020). In particular, Erik Erikson's (1994) work has shown that identity is a fluid process, located in the individual and the local culture. In tourism research, identity has been considered as a form of self-expression or self-actualisation: the type of holiday a person chooses portrays a sense of 'Self' to the outside world. Victor Turner (1969) argues that tourism can also be a 'rite of passage' as the tourist enters a new world that does not contain the constraints of everyday life.

Confronting new environments and cultures will afford the tourist opportunities for transformation and personal growth (Franklin 2003). However, tourism itself also contributes to shaping national identities. In fact, Franklin (*ibid*) argues that a nation-state actively uses tourism to construct a national narrative through its monuments, landscapes, museums and traditions. A case in point is the GDR (East Germany), which, as a newly founded state, needed to establish an 'imagined community' (Anderson 2006) in order to separate itself from the previously shared history with the FRG (West Germany). Seeing itself as the first socialist 'workers' state' on German soil, travel literature promoted museums dedicated to communist workers and the concentration camp memorials as evidence of the successful 'antifascist' fight (e.g. Autorenkollektiv 1973). Hence, heritage sites not only support the foundation of a nation-state but also influence how the state's sense of 'Self' is projected to the 'Other'. Consequently, at sites such as concentration camp memorials, visitors will have to interrogate their own positioning to the past that is presented to them.

1.1 Being a tourist at memorial sites

Chris Keil (2005) notes that the visitor experience at Auschwitz-Birkenau has clear edges as memory is framed and displaced by situating objects in glass cases. It subsequently creates a physical distance between the viewer and the object, rendering the tourist 'a semi-passive, semi-leisured and comfortable consumer' (*ibid,* 486). This view of the tourist as passive has a long tradition in tourism research, and it was not until Soile Veijola's and Eeva Jokinen's (1994) article that concepts like embodiment (e.g. emotions) entered the research in tourism. While the performative aspect of tourism plays an increasingly significant role in research, with regard to the tourist at memorial sites, it remains underresearched. Tourists are either condemned as voyeurs, as when Cole (1999) claims that visiting Auschwitz-Birkenau is the ultimate rubberneck's experience, or viewed as pilgrims. Such viewpoints create the strict pilgrim/tourist dichotomy, which most tourism researchers now reject (Olsen and Timothy 2006). Suggesting that pilgrims are not tourists ignores the fact that they also engage in typical tourist activities such as sightseeing, booking hotels and buying souvenirs. Equally, tourists can display spiritual motivations when travelling. Thus, Victor Turner and Edith Turner (1978, 20) suggest that 'a tourist is half a pilgrim, if a pilgrim is half a tourist'.

By contrast, in memory studies, the tourist is viewed as a secondary witness. For instance, Sara Jones (2014) explains in her analysis of the Stasi prison memorial Hohenschönhausen that the visitors' physical movement through the auratic space by encountering 'original' objects is a genuinely affective experience. As visitors are encouraged to identify with the victim, 'the tourist is no longer a tourist, but participant in witnessing' (*ibid,* 117). Patrizia Violi (2012) points out that a memorial visit requires emotional investment to navigate between the past and the present. Thus, the visitor at a memorial can be a witness to the traces and therefore an active participant in meaning-making processes onsite.

Viewing the tourist purely as a spectator denies him/her the emotional involvement that visitors often display, as I will demonstrate in Chapter 4 in this book. Nevertheless, Jones' argument about a site's ability to provoke genuine emotional responses oversimplifies the complexity of memorial site visits.

The cultural analyst Mieke Bal (2010) speaks of the 'grammar of a museum' when analysing exhibitions. Museums create a narrative with a beginning and an end that the visitor will follow. This narrative, however, can be influenced by the interaction between the visitor and the object. An object highlighted through lighting might draw in the visitor, while an exhibit in a less visible corner might be overlooked. Visitors will respond to exhibitions emotionally and intellectually, and curators can 'exploit' the senses to reinforce messages. At the concentration camp memorial Flossenbürg, the curatorial team decided to place the harrowing images and video footages of the death marches strategically at the end of the exhibition. Whilst the visitor can choose not to look, s/he is unable to avoid this section. The aim here is to challenge the discourse of the innocent German bystanders and to emphasise that it was indeed the Czech population that photographed the death marches. On the other hand, embedding objects within extensive sets of information can remove the visitor from the object, thus preventing emotional reactions (Dudley 2010). Consequently, when analysing visitors to memorial sites, the 'grammar' of the museum space, or indeed the wider memorial landscape, needs to be considered.

In Germany, visitor research projects are limited, partly out of a genuine fear of what the researcher might find and partly out of a sense of paralysis, because the established research methodologies appear ill-suited to the task. Volkhard Knigge (2004), the former head of Buchenwald Concentration Camp memorial, argued that one cannot transfer traditional museum research methodologies onto memorial sites, underpinning his statement with a variety of examples. He explains that, on the one hand, he is confronted with a visitor who screams at him for using gas weed killers, whilst another laments the lack of a gas chamber. On one evening he stumbled across a visitor who had missed the last bus to the nearby town of Weimar yet refused the offer of a lift as s/he felt that struggling uphill and downhill to Weimar was the only way to acknowledge the suffering of the victims. Another visitor complained about Buchenwald's ostensible lack of reconciliation as evidenced by not including Russian translations in the exhibition. How, asks Knigge, can one measure such diverse experiences with a standard visitor survey? Similarly, Stefan Küblböck (2012) suggests that visitor research at memorial sites is so complex that it is beyond the skillset of most social scientists (including himself) to conduct this research. He emphasises that any empirical research would need to include experts from different academic disciplines (e.g. psychology) in order to capture the multifaceted visitor experiences. Indeed, Michał Bilewicz and Adrian Wojcik's (2018) psychological research with over 800 high school students in Poland revealed that a visit to Auschwitz can have a profound emotional impact, to the extent that students displayed symptoms of post-traumatic stress after the visit.

Like the research in Poland, in Germany most current visitor research focuses on the school visitor. Bert Pampel (2007) finds only one study that involves observations of visitor behaviour and that not one single memorial site in Germany currently has information about its overarching visitor demographics. Moreover, only two small visitor research projects at memorial sites which commemorate the communist past had taken place prior to his research at Bautzen II. Thus, there is even less knowledge about visitors at the memorial sites that commemorate the more recent GDR past. Although Pampel's research was conducted in 2007, Habbo Knoch confirms in 2018 that visitor research at the German memorial sites remains rudimentary. In addition, a large-scale analysis of visitor comment books has so far not taken place in Germany. Consequently, the knowledge about visitors at German memorial sites is limited and tends to focus on the effectiveness of the educational programmes rather than on the individual experience of the site. Christian Gudehus (2004), for instance, points out that most surveys include questions about knowledge gained during the visit and/or the influence of the visit on personal behaviour. As such, they encourage visitors to provide the 'correct' answers instead of attempting to understand the visitor's personal perception of the site. Yet such an approach is problematic since some visitors might need time to process the visit; thus, learning effects might emerge over time rather than immediately after the visit.

The first significant visitor study in Germany, albeit with school visitors, was carried out by Dr Herbert Hötte at Neuengamme Concentration Camp memorial in 1984. This study revealed that most students were disappointed by the lack of original buildings onsite, with one of the most frequently asked questions being 'Where is the gas chamber?'. Hötte (1984) notes that, for those young people, concentration camps were synonymous with the extermination of the European Jews. The students also had only a vague knowledge about the perpetrators, as most of them made a select few high-ranking officers responsible for the Nazi atrocities; 22 per cent even stated that the Nazi regime had positive sides.

Although education was the most significant area of work for the memorial sites in East Germany, even here visitor research did not take place until the first major longitudinal study with school visitors in 1989, initiated by the Institute for Youth Research in Leipzig. The aim of this study was to gain an understanding of the impact of a visit to the Buchenwald Concentration Camp memorial by interviewing students prior to, immediately after and again five months after the visit. This research is often dismissed as being clouded by the GDR's political ideologies, since some questions were designed to capture the success of the GDR's 'antifascist narrative'. However, the most problematic aspect of this research is that it was conducted at the height of the political change in the GDR from August 1989 to December 1989, which would no doubt have influenced the students' responses.

Nevertheless, the study provides a glimpse into the students' expectations of the visit. The young East Germans often described strong emotional reactions when faced with images and/or locations of mass murder, but a number of students also reported that the site was boring as there were hardly any original features (Schubarth 1990). The research also showed that the students had only a patchy knowledge about fascism and the perpetrators. When asked several months after the visit to describe their attitude towards the Nazi regime, between 10 and 15 per cent of respondents argued that there were positive aspects of the Nazi regime. Wilfried Schubarth (*ibid*) concluded that students were increasingly desensitised and often expected to see portrayals of cruelty during their visit at a memorial site. In fact, Egon Litschke (1987), Ravensbrück's memorial manager in the GDR in the 1980s, confirmed this argument by commenting in the GDR's museum journal in 1987 that visitors increasingly remarked that the site looked 'too modern'. Thus, while the two German states differed significantly in the way they interpreted and presented the Nazi past, by the 1980s, based on these two studies, their younger generations showed an increasing emotional distance from this past as well as blind spots with regard to the perpetrators.

How memory narratives influence students' perceptions of memorial sites is poignantly demonstrated in Christian Kuchler's (2021) analysis of West German school trips to Auschwitz between 1980 and 2019. In the early 1980s, excursions to Poland involved a stop at Auschwitz; yet it was merely a part of a busy itinerary. A review of travel reports revealed that Auschwitz did not make a special impression. If students mentioned Auschwitz, then they often referred to the weather conditions, for instance rainy weather was thought to match the occasion of visiting the site. With a stronger focus on the Holocaust in West Germany in the late 1980s, Auschwitz turned into an extension of German *Vergangenheitsbewältigung* (coming to terms with the past), and the site became an important part of school trips to Poland. Students reported being overwhelmed by the hair, the glasses and the suitcases in Block five. In fact, travel reports show that some pupils broke down so that they had to leave the site. Thus, the pupils appeared to be unprepared for the shock pedagogy used in Poland at the time, a method that had already been considered inappropriate in West Germany. Moreover, shock overlay any historical learning, and thus, as Kuchler (*ibid*) notes, educational insights into the Nazi past would have been very limited. An analysis of more recent journeys to Auschwitz has, however, shown that the hair, shoes or suitcase has no longer such a prominent role in the reports. Instead, supposedly authentic structures (e.g. the *Arbeit macht frei* gate) seem to become more important. Although Kuchler's (*ibid*) research focuses on school trips, it reiterates Confino's (1997) assertion that we need to consider tourism when analysing the contemporary meaning of memory.

Since German reunification, a variety of new visitor research projects have focused predominantly on the experiences of school visitors, with the exception of a large-scale empirical research project at the concentration camp memorial Dachau in 1999. The main aim of this project was to gain information about visitor perceptions of the site in light of the proposed redesign of the exhibitions. Despite the large sample size (21,650), the research's focus on the clarity of the interpretation panels provides very little information about the visitor experience onsite (Fröhlich and Zebisch 2000). In fact, most international visitors commented that they could not answer these questions as the interpretation panels were often written only in German. Considering that 57 per cent of the visitors surveyed in this study were from abroad, the validity of the research with regard to exhibition design is therefore limited. Nevertheless, the study provides important insights into the motivations and expectations of visitors. Most visitors (22 per cent) stated that 'one has to have seen a concentration camp memorial at least once', 'I wanted to see what a concentration camp was like' (16 per cent) and 'I am interested in the history of National Socialism (14 per cent)' as the main motivational factors. Interestingly, a 'tourist' motivation was a separate category in this study (9 per cent of visitors), which emphasises Germany's distinction between the 'serious' traveller and the 'casual' tourist. Visitors mostly expected to gain an impression of the life of the prisoners (53 per cent), an understanding of people's ability to commit atrocities (46 per cent) and an increase in knowledge about the Nazi regime (44 per cent). Overall, the researchers admitted that the research process itself was challenging as most visitors were too psychologically exhausted to take part in a survey after the visit.

Pampel (2007) tried to break with the tradition of quantitative research at memorial sites during his visitor research at Bautzen II Stasi prison memorial, Münchener Platz Dresden (Nazi prison) and Pirna-Sonnenstein (Euthanasia site during the Nazi regime). His approach was to gather information about the way people processed the visit by interviewing visitors one to four years after their visit. While his sample number was low (28), the qualitative nature of his research revealed for the first time the sensory, cognitive and emotional experiences of a visit to a memorial site. Pampel (2007) concluded that visitors' experiences at memorial sites consist of five different components: emotional, cognitive, imaginative, empathetic, mnemonic, social and spatial. Visitors will review their own knowledge; they interact with space; they recall their own memories and/or family narratives; they attempt to step into the shoes of the victims and/or perpetrators; and they develop mental images by engaging with the narrative that is represented. He further emphasises that the visitor's response to the memorial site is a highly individual one, influenced by the visitor's cultural background. In addition, unlike school visitors, motivations for visiting are intrinsic, meaning that visitors can name a specific reason for their visit. Pampel's research confirms the anthropological perspective of tourism. For the individual tourist, the immediacy of the historical space is the most important aspect of the visit; learning is a by-product.

Another recently conducted visitor study analysed TripAdvisor comments at two very different sites related to the Holocaust: the Otto Weidt Workshop for the Blind (Otto Weidt rescued blind Jews from deportation during the Second World War) and the Jewish Museum in Berlin (Souto 2018). The analysis has shown that the personal stories rather than the objects were the most memorable experiences for visitors. At the Jewish museum in Berlin, visitors also highlighted the experiential architecture of the Tower of Holocaust, designed to disorientate visitors and turn them into active participants in memory work. Although the Otto Weidt Workshop is rather plain in comparison to the Jewish Museum Berlin, the reaction was often of a similar emotional nature. For example, one visitor remarked: 'The museum/exhibition was so touching and thought provoking, I must admit to having a lump in my throat and could have easily got emotional whilst there' (*ibid*, 22). Several other visitors also commented on the 'rough walls' or the fact that 'one cannot tell that they [the rooms] have been renovated at all'. Souto's and Pampel's research have shown that visitors attempt a form of empathetic matching, which does not necessarily require modern interpretative museum technologies. It is the grammar of a museum that evokes such reactions.

All of the research projects mentioned earlier, however, have in common the analysis of visitors' responses after their visit to the museum and/or memorial site, which does not shine a light on the experience as and when it happens. The first larger study involving participant observations was conducted by Victoria Bishop Kendzia (2017) at the Jewish Museum in Berlin. While the focus was again on school visitors, Bishop Kendzia's research shows for the first time how visitors negotiate the museum space. Her ethnographic research revealed that pupils from the former West Germany knew how to behave in the museum and in particular in the Holocaust tower. The students displayed a solemn and sombre body language and expressed feelings of guilt when asked about their relationship to Jewish history. This corresponds with Casey's (1987) argument that emotions are stored in places and with Paul Connerton's (1989) theory that memories manifest themselves through ritualised bodily behaviour. Thus, when researching visitors at memorial sites, one needs to be aware of the cultural codes that visitors subscribe to. In stark contrast stood the experience of a school group from the former East Germany. Although they were born after German reunification, they did not display guilt, sadness, *Betroffenheit* (a form of sadness/dismay) or sympathy. Bishop Kendzia (*ibid*) concludes that these East Germans have not adopted the 'Western' Holocaust discourse, so they reacted in a more detached manner. It emphasises one of the blind spots in German memory culture and indeed academic research in this field. Germany is often praised for its way of dealing with the Nazi past since 1990; yet the underlying assumption is that the East Germans unquestionably adopted the new memory culture. In fact, Matthias Heyl (2016) argues that not only do we have little knowledge about the educational programmes at the memorial sites in the GDR, we also have little information about how these memory discourses might be passed on to future generations, which, according to Bishop Kendzia, appears to happen. Moreover, a school group with most students from a migration background (Turkish) was even scolded by museum staff for not showing the required German *Betroffenheit*; hence, Bishop Kendzia (2017) argues that memory culture in Germany is also an expression of 'Germanness', which excludes people who do not conform to those codes.

1.2 Performing authenticity

Visitors to Dachau often bemoan on TripAdvisor the underwhelming emotional experience caused by the large visitor numbers who disrupt the 'sombre' atmosphere. Considering this comment, the notion of authenticity seems to go beyond the material evidence. Yet museums at memorial sites often focus on objects that represent the suffering of the past. In fact, Jens-Christian Wagner (2017, 8) argues that most museum curators at concentration camp memorials 'cling onto museum objects for dear life' in an attempt to explain the Nazi past. In so doing, the atrocities are reduced to key artefacts. Their repeated reproduction, however, seems to destroy the authentic aura visitors seek. Thomas Thurnell-Read (2009) established during a research project with young travellers at Auschwitz that the striped uniforms in glass cases, although historically authentic, did not provoke any emotional reaction. Indeed, they seem to create a sense of unreality and anti-climax. In addition, these objects are subject to decay, and decisions need to be made about what can (and cannot) be preserved. The hair in the glass cases at Auschwitz is slowly turning to grey dust, and the barbed wire fences are rusting. Should the hair be treated chemically in order to preserve it, and if so, is it still then 'authentic'?

The concept of authenticity is part of a central debate in tourism research which has produced two strands of thinking: 'object-related authenticity' and 'existential authenticity' linked to emotions (Knudsen and Waade 2010). MacCannell (1973, 594) was the first sociologist to investigate the tourist's desire 'to join in with the real life'. He introduced the concept of staged authenticity, that is, the tourist destination creates an authentic setting depending on what the tourist would believe to be authentic. He claimed that tourists are usually unable to experience the 'real' destination as they are not allowed to look behind the scenes; they are sheltered from real life so that their dream of an exciting holiday is fulfilled.

Yet MacCannell (*ibid*) ignored the intense feelings that places – in particular memorial sites – can evoke. Hence, Ning Wang (1999) launched the term 'existential authenticity'. He (*ibid*, 352) describes authenticity as a 'state of being', derived from the psychological theories of existentialism. Unlike MacCannell, Wang does not link authenticity to objects. Indeed, Pirker et al. (2015) emphasise that the original object is no longer sufficient in a postmodern society; it is the narrative surrounding the object which is essential to experience authenticity. Subsequently, Pirker et al. (*ibid*) differentiate between two authentic modes: *authentisches Zeugnis* (authentic witness) and *authentisches Erleben* (authentic experience). An authentic witness is characterised by its genuineness, whilst an authentic experience is achieved through a provocation of feeling (e.g. re-enactments, development of an authentic atmosphere).

The provocation of feelings is crucial to the concept of authenticity at memorial sites. For a visitor to have a sense of authenticity, s/he needs to use the power of imagination. The sites are characterised by absence, since the dead who would be witnesses to the atrocities cannot have a voice. Hence, the truthfulness of the traces is not sufficient; it is the supposed authenticity as the link between the past and the present. Moreover, authenticity is based on what the visitor brings to the occasion. If a visitor has substantial knowledge about the atrocities committed onsite, s/he is likely to regard the site as an 'authentic' experience. Thus, 'the authenticity effect is a meaning effect produced in the experience of the visitor' (Violi 2012, 44) rather than an abstract form of reality. Britta Knudsen and Anne Waade (2010), therefore, introduce the notion of 'indexical authenticity', emphasising the performative aspect of authenticity. They link the intense feelings a tourist experiences to the referential character of a site. It is the proximity to the event that creates authentic feelings. Hence, for the first time, Knudsen and Waade combine object authenticity with existential authenticity to highlight that both are required for a tourist to experience a site as truly authentic.

There is no better example to demonstrate Knudsen's and Waade's concept of indexical authenticity than Colditz Castle. Amongst British visitors, Colditz is famously known as the former prisoner-of-war (POW) camp Oflag IVc. They usually have substantial knowledge about the site prior to arrival, largely derived from popular media representations such as *The Colditz Story* and tales of heroic British escapes.[1] Thus, the 'authenticity effect' is created by being in the space and seeing the traces of the former POW camp. However, this stands in contrast to the German visitors, who are likely to have little knowledge about the POW camp and for whom consequently the site had no meaning and so is not an authentic experience. Colditz also points to another argument raised by Assmann (2011): the loss of authenticity that comes from turning the site into a memorial and/or museum. The chapel at Colditz, untouched since its liberation in 1945, was a scene of great awe for British visitors. Its dilapidated state allowed them to 'time travel' to the period of the POW camp. Yet the chapel was structurally unsafe and would have needed to be restored to prevent further decay. The paradox is that any restorative intervention would reduce the feelings of authenticity.

For memorial sites, this is a significant dilemma. The visitor expects a site 'frozen' in time, but memorials, and in particular former concentration camps, were never built to last. They require, as do many historical buildings, substantial maintenance to prevent decay. However, restoration can destroy the referential character of the past trauma for today's tourists, since a dilapidated condition creates the 'meaning effect' for them. Moreover, memorial sites also have to strike a balance between an 'authentic' representation and meeting the needs of the eyewitnesses. At Bautzen II, former prisoners requested the management team to place soft toy rats in the basement of the prison in order to demonstrate the poor working conditions prisoners were subjected to. This idea was rejected by the management team as the rats would not have contributed to a greater historical understanding on the visitors' part (Klewin 2016, personal

---

1 Unpublished dissertation: Pastor, D. (2007) – Dark Tourism? – Eine Besucheranalyse am Fallbeispiel Schloss Colditz, Hochschule Harz, Fachbereich Wirtschaftswissenschaften.

communication). This example poses additional questions. Who determines 'authenticity' and therefore who sets the agenda in memorial museums? Is it the survivor, the curator or indeed the visitor? During the set-up phase of a memorial site, authenticity is a negotiation between the curatorial team and the eyewitnesses. As time passes, the authenticity shifts into 'meaning-making' for the visitor. Yet the current German memorial concept refers to authenticity only with regard to material authenticity, creating a discrepancy between institutional aims and visitor expectations.

In light of the complexity of tourism to memorial sites, Duncan Light (2017) urged academics to get close to tourists to gain an understanding of the rising phenomenon of visiting sites of atrocities and to engage with neighbouring disciplines like memory studies with a view to shining a light on identity-formation processes. Similarly, Alena Pfoser (2017) highlights that the dynamics of memory and their circulation among tourists is an important area of investigation. For anthropologists, a major aim is to understand the meaning tourists create during their travels (Graburn 2002). While such research does not provide any information about the tourist's contribution to the local economy or the likelihood of them returning, it will reveal an in-depth account of a tourist's inner and outer journey (*ibid*, 30). Since a visit to highly symbolic environments involves 'an inner movement of construction of meaning and an outer movement of encounters with sensual and symbolic objects' Zachary Beckstead (2010, 390) suggests looking for catalysts which enable heightened emotions. Anthropologists often share the frustration that the voices of the tourists are very rarely heard as one often talks *about* them but not *with* them (Bruner 2005; Harrison 2002). Hence, the key aim of this book is to give a voice to the tourist who visits memorial sites.

1.3 Structure of the book

In Chapter 2, I will provide a brief overview of Germany's specific approach to memorialisation. It was not until reunification in 1990 that Germany could confront the Nazi past as a unified nation state. Prior to that, the two German states, the FRG and the GDR, pursued different strategies; these have an impact on the memorial sites and indeed on the visitor to this day. I will also explore the concept of *Gedenkstättenpädagogik* (education at memorial sites) as it guides the design of the exhibitions and the management of the sites. Aside from confronting the Nazi past, contemporary Germany is also committed to coming to terms with the second dictatorship of the GDR. Whilst the focus in the 1990s was the Stasi and its oppressive nature, in more recent years the GDR's everyday culture came to the fore. Moreover, the opening of the *Treuhand* archives (the organisation which was responsible for the privatisation of GDR's publicly owned companies) in 2018 led to renewed debates about the GDR and its aftermath. Hence, memory debates about the GDR legacy are ongoing and at times still heated.

In Chapter 3, I will explore the development of the memorial sites that underpinned this research. German memory politics is imprinted onto concentration camp memorials Ravensbrück and Flossenbürg, with both sites largely characterised by absence. Most of Ravensbrück was transformed into a Soviet army base after 1945, which left its mark on the site. After German reunification, the management team faced the arduous task of redeveloping the site according to (West) Germany's concept of memorialisation. Buildings that interfered with the memory of the concentration camp were demolished, while the former area of the barracks was covered with a surface made of clinker that revealed the outline of the barracks. At Flossenbürg, the local authority was keen to fend off the stigma of the former concentration camp by transforming the site into an industrial and housing estate since employment and accommodation was urgently needed for the refugees from the Eastern territories, thus creating a multi-layered memoryscape. The House of the Wannsee Conference escaped destruction of historical evidence, yet its significance as the site of the infamous 'final solution' Conference was largely forgotten. Several unsuccessful attempts to open a memorial delayed the opening of the site until 1992. Since then, exhibitions have changed three times, signifying Germany's struggle with exhibition design at a perpetrator site. Unlike the memorial sites that commemorate the Nazi past, memorials that deal with the GDR were established swiftly after the fall of the Berlin Wall. Bautzen II, opened in 1994, struggles nevertheless with exhibiting the complex past of the Soviet Special Camps, Nazi imprisonment and the Stasi in one confined space.

In Chapter 4, I will analyse the visitor experiences at German memorial sites that were used in my research. At Flossenbürg and Ravensbrück, authentic structures can cause intense reactions, while at the same time exhibitions can feel repetitive. In fact, the landscape design at the memorial sites is as important as the exhibitions, if not more so. In addition, visitors approach these sites with a significant amount of 'baggage', which contains the cultural background, the family narrative and the national discourse. The House of the Wannsee Conference emphasised how influential national memory discourses are. Although Wannsee is a former perpetrator site, visitors from Israel approached it as victims and with a sense of pride. By contrast, German visitors often felt ashamed and struggled with family members' involvement in the Nazi regime. There was, however, also a large group of visitors for whom the House of the Wannsee Conference was the location of the film *Conspiracy* and must therefore be seen while being in Berlin. Bautzen II stands out as the site with the most intense reactions. On the one hand, visitors struggled to cope with the prison atmosphere, while on the other hand visitors showed signs of sensationalism, comparing the site to other prisons. The memorial sites also showed the constructed nature of authenticity. While at the concentration camp memorials visitors bemoaned at times the lack of historical evidence, at Bautzen visitors commented that one would never be able to exhibit what happened onsite. Therefore, Bautzen also revealed the challenges of memory tourism in post-conflict societies, which, if not carefully managed, can entrench existing hostilities.

In Chapter 5, I reflect on memory tourism more broadly. While research into dark tourism shines a light on visitor motivations and the challenges of managing 'dark' sites, the ethnographic mode of this research emphasised the need for a wider view of tourism to memorial sites. The tourist is not a passive consumer; s/he is an active participant in memory processes. In fact, the sites are part of an identity-formation process since visitors have to position themselves in relation to the past that is exhibited. Moreover, senses can trigger embodied emotions which are foregrounded in pre-existing knowledge. Thus, atmosphere and historical knowledge form a powerful bond that turns the tourist into an experiential visitor. Emotions at memorial sites are therefore complex, and empathy is not a universal outcome. Indeed, I argue that, while a memorial museum might plan for an empathetic reaction they can, however, not predetermine it as an outcome of the visit. In line with this, the notion of memorial sites possessing transformative powers requires a review. If visitors' motivation to visit a memorial site is born out of intrinsic values, then memorial sites can be spaces of critical thinking. If, however, the motivation is to chase symbolic representations of the past, then the potential for a transformation is limited. Visitors are also often criticised for staging inappropriate photographs. This research revealed that German visitors experience a cultural barrier; hence, instead of universally condemning the taking of photographs, an analysis of the frame is required, that is who/what is in the photo.

Chapter 6 considers the future of German memorial sites. Germany's memory culture, although often celebrated, is at a critical juncture. With the death of the last Holocaust survivors, the era of the *Zeitzeugen* (similar to the English word eyewitnesses) is coming to an end, requiring new formats of engagement. In an attempt to 'never forget', ever-new initiatives such as the virtual Holocaust survivor are developed. Yet we know very little about the impact 'digital memory' has on actual memory. Similarly, Germany's much more diverse society, often with their own traumatic memories, demands a critical review of existing pedagogical approaches. I, therefore, suggest in this chapter the democratisation of memorial sites by, for instance, including 'ordinary' people in the management of the sites, rather than the top-down approach from 'experts' that we currently see. More opportunities for participation might also close the divide between the local community and the memorial site in contested places like Bautzen. And finally I argue that society needs to take a critical look at current methodologies of exhibiting 'difficult pasts' in museums, which might encourage us to remember the past but does not necessarily lead to active, ethical citizenship.

1.4 Bibliography

Anderson, Benedict. 2006. *Imagined Communities: Reflections on the Origin and Spread of Nationalism*. London; New York: Verso Books.

Assmann, Aleida. 2011. *Erinnerungsräume: Formen und Wandlungen des kulturellen Gedächtnisses*. 5th ed. München: C.H. Beck.

Autorenkollektiv. 1973. *Reiseführer Deutsche Demokratische Republik*. Leipzig: Brockhaus.

Bal, Mieke. 2010. 'Guest Column: Exhibition Practices'. *PMLA* 125 (1): 9–23.

Beckstead, Zachary. 2010. 'Commentary: Liminality in Acculturation and Pilgrimage: When Movement Becomes Meaningful'. *Culture & Psychology* 16 (3): 383–393.

Bilewicz, Michał, and Adrian Dominic Wojcik. 2018. 'Visiting Auschwitz: Evidence of Secondary Traumatization among High School Students'. *American Journal of Orthopsychiatry* 88 (3): 328–334.

Bishop Kendzia, Victoria. 2017. *Visitors to the House of Memory: Identity and Political Education at the Jewish Museum Berlin*. New York: Berghahn.

Bruner, Edward M. 2005. *Culture on Tour*. Chicago: The University of Chicago Press.

Casey, Edward S. 1987. *Remembering: A Phenomenological Study*. Bloomington: Indiana University Press.

Cattell, Maria G., and Jacob J. Climo. 2002. 'Meaning in Social Memory and History: Anthropological Perspectives'. In *Social Memory and History: Anthropological Perspectives*, edited by Jacob J. Climo and Maria G. Cattell, 1–38. Walnut Creek: Altamira Press.

Cole, Tim. 1999. *Selling the Holocaust: From Auschwitz to Schindler: How History Is Bought, Packaged, and Sold*. New York: Psychology Press.

Confino, Alon. 1997. 'Collective Memory and Cultural History: Problems of Method'. *The American Historical Review* 102 (5): 1386–1402.

Connerton, Paul. 1989. *How Societies Remember*. Cambridge: Cambridge University Press.

Culler, Jonathan. 1981. 'Semiotics of Tourism'. *American Journal of Semiotics* 1 (1): 127–140.

Cywiński, Piotr M.A. 2019. 'Memory of Tomorrow'. *Observing Memories*: 18–21, 3.

Dudley, Sandra H. 2010. 'Museum Materialities: Objects, Sense and Feeling'. In *Museum Materialities: Objects, Engagement, Interpretations*, edited by Sandra H. Dudley, 1–18. London; New York: Routledge.

Einhorn, Barbara. 2000. 'Gender, Nation, Landscape and Identity in Narratives of Exile and Return'. *Women's Studies International Forum* 23 (6): 701–713.

Erikson, Erik H. 1994. *Identity and the Life Cycle*. New York: Norton & Company.

Fentress, James, and Chris Wickham. 2012. *Social Memory*. New York: American Council of Learned Societies.

Franklin, Adrian. 2003. *Tourism: An Introduction*. London: Sage Publications.

Fröhlich, Werner, and Johanna Zebisch. 2000. 'Besucherbefragung zur Neukonzeption der KZ-Gedenkstätte Dachau. Ergebnisbericht'. www.hdbg.de/basis/pdfs/besucherforschung_neugestaltung-der-ausstellung-in-der-kz-gedenkstaette-dachau.pdf.

Fussell, Paul. 1982. *Abroad: British Literary Traveling between the Wars*. Oxford: Oxford University Press.

Gable, Eric, and Richard Handler. 2000. 'Public History, Private Memory: Notes from the Ethnography of Colonial Williamsburg, Virginia, USA'. *Ethnos* 65 (2): 237–252.

Geertz, Clifford. 1973. *The Interpretation of Cultures*. New York: Basic Books.

Graburn, Nelson. 2002. 'The Ethnographic Tourist'. In *The Tourist as a Metaphor of the Social World*, edited by Graham Dann, 19–40. Wallingford: CABI Publishing.

Gravari-Barbas, Maria, and Nelson Graburn. 2012. 'Tourist Imaginaries'. *Via* 2012 (1): 1–7.

Gudehus, Christian. 2004. 'Methodische Überlegungen zu einer Wirkungsforschung in Gedenkstätten'. In *Lagersystem und Repräsentation. Interdisziplinäre Studien zur Geschichte der Konzentrationslager*, edited by Ralf Gabriel, Elissa Mailänder-Koslov, Monika Neuhofer, and Else Rieger, 206–219. Tübingen: Edition Diskord.

Halbwachs, Maurice. 1925. *Les Cadres Sociaux de La Mémoire*. Paris: Felix Alcan.

Hansen-Glucklich, Jennifer. 2014. *Holocaust Memory Reframed: Museums and the Challenges of Representation*. New Brunswick, NJ: Rutgers University Press.

Harrison, Julia. 2002. *Being a Tourist: Finding Meaning in Pleasure Travel*. Vancouver, Canada: UBC Press.

Hertzog, Anne. 2012. 'War Battlefields, Tourism and Imagination'. *Via. Tourism Review* 1 (March). https://doi.org/10.4000/viatourism.1283.

Hertzog, Anne. 2019. 'Tourism and Places of Memory: Exploring the Political Side of Tourism and the Spatial Dimension of Memory'. *Observing Memories* 3: 6–17.

Heyl, M. 2016. Personal conversation. *Exhibition Design at Ravensbrück*, 20 July 2016.

Hötte, Herbert. 1984. 'Museumspädagogische Arbeit mit Jugendlichen im Dokumenten-haus KZ Neuengamme'. *Internationale Schulbuchforschung* 6 (2): 173–185.

Hunt, Nigel, and Sue McHale. 2007. 'Memory and Meaning: Individual and Social Aspects of Memory Narratives'. *Journal of Loss and Trauma* 13 (1): 42–58.

Jones, Sara. 2014. *The Media of Testimony: Remembering the East German Stasi in the Berlin Republic*. Basingstoke: Palgrave MacMillan.

Keats, Patrice A. 2005. 'Vicarious Witnessing in European Concentration Camps: Imagining the Trauma of Another'. *Traumatology* 11 (3): 172–183.

Keil, Chris. 2005. 'Sightseeing in the Mansions of the Dead'. *Social and Cultural Geography* 6 (4): 479–494.

Kermani, Navid. 2018. *Entlang den Gräben: Eine Reise durch das östliche Europa bis nach Isfahan*. Munich: C.H. Beck.

Klewin, S. 2016. Personal conversation. *The Management of the Memorial Site Bautzen II*, 20 September 2016.

Klüger, Ruth. 2001. *Still Alive: A Holocaust Girlhood Remembered*. New York: Feminist Press at the City University of New York.

Knigge, Volkhard. 2004. 'Museum oder Schädelstätte? Gedenkstätten als multiple Institutionen'. In *Museumsfragen. Gedenkstätten und Besucherforschung*, edited by Stiftung Haus der Geschichte der Bundesrepublik Deutschland, 17–33. Bonn: Stiftung Haus der Geschichte der Bundesrepublik Deutschland.

Knoch, Habbo. 2018. 'Gedenkstätten'. *ZZF – Centre for Contemporary History*. https://doi.org/10.14765/zzf.dok.2.1221.v1.

Knudsen, Britta Tim, and Anne Marrit. 2010. 'Performative Authenticity in Tourism and Spatial Experience: Rethinking the Relations between, Travel, Place and Emotions'. In *Re-Investing Authenticity: Tourism, Place and Emotions*, 1–22. Bristol: Channel View Publications.

Küblböck, Stefan. 2012. 'Sich selbst an dunklen Orten begegnen: Existentielle Authentizität als Potenzial des Dark Tourism'. In *Dark Tourism: Faszination des Schreckens*, edited by Heinz-Dieter Quack and Albrecht Steinecke, 114–124. Paderborn: Selbstverl. des Faches Geographie, Fak. für Kulturwiss., Univ. Paderborn.

Kuchler, Christian. 2021. *Lernort Auschwitz: Geschichte und Rezeption schulischer Gedenkstättenfahrten 1980–2019*. Göttingen: Wallstein Verlag.

Lennon, John, and Malcolm Foley. 2010. *Dark Tourism: The Attraction of Death and Disaster*. London; New York: Cengage Learning.

Light, Duncan. 2017. 'Progress in Dark Tourism and Thanatourism Research: An Uneasy Relationship with Heritage Tourism'. *Tourism Management* 61 (August): 275–301.

Light, Duncan, Remus Crețan, and Andreea-Mihaela Dunca. 2020. 'Transitional Justice and the Political "Work" of Domestic Tourism'. *Current Issues in Tourism*. https://doi.org/10.1080/13683500.2020.1763268.

Lisle, Debbie. 2016. *Holidays in the Danger Zone*. Minneapolis: Minnesota University Press.

Litschke, Egon. 1987. 'Ehemaliger Zellenbau in der Nationalen Mahn- und Gedenkstätte Ravensbrück rekonstruiert und umgestaltet'. *Neue Museumskunde* 87 (4): 308–310.

MacCannell, Dean. 1973. 'Staged Authenticity: Arrangements of Social Space in Tourist Settings'. *American Journal of Sociology* 79 (3): 589–603.

———. 1989. *The Tourist: A New Theory of the Leisure Class*. Berkeley; Los Angeles: University of California Press.

Malpass, Jeff. 2011. 'Place and the Problem of Landscape'. In *The Place of Landscape: Concepts, Contexts, Studies*, edited by Jeff Malpass, 3–26. Cambridge, MA: MIT Press.

Moller, Sabine. 2010. 'Das kollektive Gedächtnis'. In *Gedächtnis und Erinnerung. Ein interdisziplinäres Handbuch*, edited by Christian Gudehus, Ariane Eichenberg, and Harald Welzer, 85–94. Stuttgart: J.B. Metzler.

Niedenthal, Paula M., and Marcus Maringer. 2009. 'Embodied Emotion Considered'. *Emotion Review* 1 (2): 122–128.

Olsen, Daniel H., and Dallen J. Timothy. 2006. 'Tourism and Religious Journeys'. In *Tourism, Religion and Spiritual Journeys*, edited by Dallen J. Timothy and Daniel H. Olsen, 1–22. London; New York: Routledge.

Pampel, Bert. 2007. *'Mit eigenen Augen sehen, wozu der Mensch fähig ist': Zur Wirkung von Gedenkstätten auf ihre Besucher*. Frankfurt am Main: Campus Verlag.

Pfoser, Alena. 2017. 'Tourism and Transnational Memory Formation in Tallinn, Estonia'. 6 (April). www.europenowjournal.org/2017/04/03/tourism-and-transnational-memory-formation-in-tallinn-estonia/.

Pirker, Eva Ulrike, Mark Rüdiger, Christa Klein, Thorsten Leiendecker, Carolyn Oesterle, Miriam Sénécheau, and Michiko Uike-Bormann. 2015. *Echte Geschichte: Authentizitätsfiktionen in populären Geschichtskulturen*. Bielefeld: Transcript Verlag.

Pollock, Griselda. 2003. 'Holocaust Tourism: Being There, Looking Back and the Ethics of Spatial Memory'. In *Visual Culture and Tourism*, edited by Nina Lubbren and David Crouch, 175–190. Oxford: Berghahn.

Popescu, Diana I. 2016. 'Post-Witnessing the Concentration Camps: Paul Auster's and Angela Morgan Cutler's Investigative and Imaginative Encounters with Sites of Mass Murder'. *Holocaust Studies* 22 (2–3): 274–288.

Rickly-Boyd, Jillian M. 2012. 'Authenticity & Aura'. *Annals of Tourism Research* 39 (1): 269–289.

Robinson, Joseph. 2019. *Transitional Justice and the Politics of Inscription: Memory, Space and Narrative in Northern Ireland*. London: Routledge.

Schubarth, Wilfried. 1990. *Wirkungen eines Gedenkstättenbesuches bei Jugendlichen: (Ergebnisse einer Wirkungsanalyse von Besuchen in der Nationalen Mahn- und Gedenkstätte Buchenwald)*. https://nbn-resolving.org/urn:nbn:de:0168-ssoar-403823.

Schult, Tanja. 2018. 'Introduction (to the Special Issue Performative Commemoration of Painful Pasts)'. *Liminalities: A Journal of Performance Studies* 14 (3): 3–11.

Souto, Ana. 2018. 'Experiencing Memory Museums in Berlin: The Otto Weidt Workshop for the Blind Museum and the Jewish Museum Berlin'. *Museum and Society* 16 (1): 1–27.

Stone, Philip R. 2011. 'Dark Tourism: Towards a New Post-Disciplinary Research Agenda'. *International Journal of Tourism Anthropology* 1 (3/4): 318–332.

Sumartojo, Shanti. 2016. 'Commemorative Atmospheres: Memorial Sites, Collective Events and the Experience of National Identity'. *Transactions of the Institute of British Geographers* 41 (4): 541–553.

———. 2019. 'Sensory Impact: Memory, Affect and Sensory Ethnography at Official Memory Sites'. In *Doing Memory Research: New Methods and Approaches*, edited by Danielle Drozdzewski and Caroline Birdsall, 21–37. Singapore: Palgrave Macmillan.

Thomas, Emily. 2020. *The Meaning of Travel: Philosophers Abroad*. Oxford: Oxford University Press.

Thurnell-Read, Thomas. 2009. 'Engaging Auschwitz: An Analysis of Young Travellers' Experience of Holocaust Tourism'. *Journal of Tourism Consumption and Practice* 1 (1): 26–52.

Tischer, Stefan. 2020. Personal communication with the author, 10 January 2020.

Turner, Victor W. 1969. *The Ritual Process: Structure and Anti-Structure*. Chicago: Aldine Publishing Company.

Turner, Victor W., and Edith Turner. 1978. *Image and Pilgrimage in Christian Culture*. New York: Columbia University Press.

Veijola, Soile, and Eeva Jokinen. 1994. 'The Body in Tourism'. *Theory, Culture & Society* 11 (3): 125–151.

Violi, Patrizia. 2012. 'Trauma Site Museums and Politics of Memory: Tuol Sleng, Villa Grimaldi and the Bologna Ustica Museum'. *Theory, Culture & Society* 29 (1): 36–75.

Wagner, Jens-Christian. 2017. 'Gedenkstättenarbeit in Deutschland seit 1945: Eine Erfolgsgeschichte?'. *Netzwerk Erinnerung und Zukunft Hannover e.V.* (September): 5–10.

Wang, Ning. 1999. 'Rethinking Authenticity in Tourism Experience'. *Annals of Tourism Research* 26 (2): 349–370.

Wasserman, Judith R. 1998. 'To Trace the Shifting Sands: Community, Ritual, and the Memorial Landscape'. *Landscape Journal* 17 (1): 42–61.

White, Geoffrey M. 1999. 'Emotional Remembering: The Pragmatics of National Memory'. *Ethos* 27 (4): 505–529

Winter, Caroline. 2015. 'Ritual, Remembrance and War: Social Memory at Tyne Cot'. *Annals of Tourism Research* 54 (September): 16–29. https://doi.org/10.1016/j.annals.2015.05.001.

Xie, Yanjun, and Jiaojiao Sun. 2018. 'How Does Embodiment Work in Dark Tourism "Field"? Based on Visitors' Experience in Memorial Hall of the Victims in Nanjing Massacre'. *International Journal of Tourism Cities* 4 (1): 110–122.

2 The institutionalisation of memory in Germany

2.1 The development of the German concentration camp memorials

Whilst in the English-speaking academic literature the terms memorial museums (Williams 2008) or memory museums (Arnold-de Simine 2012) are used to describe museums that exhibit violence, this cannot be easily transferred to the German term *Gedenkstätten*. The word entered the German language after the Second World War with the GDR introducing *Nationale Mahn-und Gedenkstätten* (national warning and memorial sites) as early as the 1950s. In the FRG (West Germany), *Gedenkstätten* developed in the 1960s with the opening of the memorial sites at Bergen-Belsen, Dachau and Neuengamme (Knoch 2018). The memorial sites' main aims were, and still are, scientific research, caring for survivors and their relatives, and the preservation of historical evidence. Hence, the sites have first and foremost a humanitarian function with the exhibitions being a side effect of the historical research. Furthermore, *Gedenkstätten* can only exist at sites with locational authenticity.

The development of *Gedenkstätten* in East and West Germany was fundamentally different. In the GDR, *Gedenkstätten* were an integral part of the political landscape. Designed as beacons of antifascism, they were the symbolic evidence of the 'better state'. The focal point in the GDR was Buchenwald near the town of Weimar, not least due to the narrative of the *Buchenwaldkind* and the death of the communist Ernst Thälmann. In 1952 the first exhibition at Buchenwald was opened in the *Torgebäude*, and in 1954 a second exhibition followed in the former canteen building. In the meantime, the memorial for Ernst Thälmann was erected near the crematorium in 1953. Thus, the communist victims took centre stage at Buchenwald. With the publication of Bruno Apitz's book *Nackt unter Wölfen* (*Naked among wolves*), which was later adapted for a film *Nackt unter Wölfen*, Buchenwald was firmly anchored in GDR memory as the site where communists rescued a Jewish child. The child, however, could only survive because another child was added to the deportation list, a fact that was silenced in the GDR (Niven 2007).

Most school visits to a concentration camp memorial, compulsory in the GDR, occurred in Buchenwald. However, Buchenwald was also promoted

DOI: 10.4324/9781003126836-2

as a major memorial site to international visitors. A leaflet of the GDR's air-line Interflug highlighted Buchenwald as a place to visit, and organised trips by the GDR's state-owned tourism authority *Reisebüro der DDR* to Weimar often included a visit to Buchenwald. In fact, even the GDR's guide to art and culture included a detailed description of Buchenwald for domestic and international visitors (e.g. Piltz 1985). Since the GDR did not view itself as the successor state of the Third Reich, it could approach the active promotion of memorial sites as part of the tourism landscape without feelings of unease. Indeed, it was part of the GDR's identity construction; a visitor in Buchenwald during the GDR era would have encountered anti-western propaganda and an emphasis on the glorious antifascist fight (Niven 2009). The memorial site of Sachsenhausen in Oranienburg enjoyed a similar fame in the GDR, while Ravensbrück, the women's concentration camp, played a minor role.

The GDR's *Mahn- und Gedenkstätten* were centrally managed by the Minis-try of Culture to ensure that education and exhibition programmes adhered to the SED's (Sozialistische Einheitspartei – socialist unity party) claim to power. In general, the focus was on the communist victims, to the detriment of all other victim groups. Exhibitions were designed to induce shock and hatred against the 'imperialist fascist system', of which West Germany was seen as a continuation. As such, the GDR's exhibitions were often designed to contrast 'good and evil', using images or objects that made an emotional impact: *Prügel-block* (a rack for flogging), broken teeth and lamp shades made of skin on the one side and heroic communist resistance fighters on the other (Leo 1998).

Since the completion of the memorial sites until the 1970s, survivors led most of the guided visits at the memorial sites. Thus, these early visits to con-centration camp memorials in the GDR could have been very emotional; yet some preliminary research has shown that most survivors stuck to the official educational guidelines with a focus on the GDR as the new antifascist state. Adhering to these official narratives would have not only avoided conflicts with the overarching body of the Committee of the Antifascist fighters but also helped to cope emotionally. It would have been psychologically impossible to recall one's own traumatic experiences during daily guided tours. Indeed, a manuscript of guided tours at Sachsenhausen Concentration Camp memorial revealed that the tours conducted by the former prisoner Ernst Harter did not contain any personal experiences. If he had not used the word 'we' occasion-ally, one would have not noticed that he was a survivor (*ibid*).

By contrast, in West Germany the first memorial sites initiated often under allied administrations remained rudimentary. Unlike in the GDR, the early memorial sites did not form part of nation-building and were predominantly sacred commemorative sites. Moreover, the sites were often re-purposed or transformed in such a way that they were unrecognisable as former concentra-tion camps, for example, Flossenbürg or Neuengamme. It was not until groups of survivors or local activists campaigned for more comprehensive memo-rial sites that exhibitions and/or educational programmes were developed at Dachau, Neuengamme, Bergen-Belsen and Flossenbürg. Looking back,

Cornelia Brink (1998) suggests that one can distinguish three developmental phases at the memorial sites in West Germany: phase one from 1945 to 1949, phase two from 1949 to 1968 and phase three from 1968 onwards. In the post-war years, memorials at former concentration camps were often initiated by the allies and/or survivors, for example, at Bergen–Belsen the British troops demanded the erection of memorial plaques. For the survivors, however, such forms of memorialisation were even in these early years not enough. Consequently, at Bergen–Belsen British soldiers created an exhibition with photographs from the camp, and in Dachau former prisoners together with US army soldiers designed an exhibition about the Dachau trials in 1945.

During the second phase from 1949 to 1968, the focus of memorialisation was on the resistance movement and the attempted plot to assassinate Hitler on 20 July 1944. Thus, the development of an exhibition at Plötzensee was emphasised, while the early exhibition in Dachau was closed and not reopened in a new format until 1960. The re-emergence of antisemitism in West Germany in the 1960s constituted a first caesura in West German memory culture. In Dachau, the first permanent exhibition and an archive was created, while *Aktion Sühnezeichen* launched a small documentation centre at Bergen–Belsen. These exhibitions, however, were not underpinned by pedagogical concepts. Their task was first and foremost to document the historical evidence and to engage with political education, not how to present Nazi crimes to a public audience. During these years, the perpetrator gaze was the dominant form of exhibition design as Nazi photographs were presented uncritically. Perpetrators were contrasted with helpless victims, while bystanders, the structure of the concentration camps and the relationship to the local community were absent.

A third phase commenced with the student movement in 1968. This generation provoked intense debates about the pervasive silence about the Nazi regime in German society and demanded a more comprehensive reckoning with the past. With a previously strong focus on the political and economical structures of the Third Reich, the focus shifted now to the society. Hence, the concentration camps became new focal points for local historians and activists that led to a *Gedenkstätten* boom. In order to distinguish these sites from ordinary museums, the activists referred to them as *Gedenk- und Informationsstätten*, thus stressing the importance of working with historical documents and survivors instead of conducting ritualised commemorative activities (Knoch 2020). In the 1980s, the term *Gedenkstätten* was increasingly used in West Germany, especially with the launch of the *Gedenkstättenreferat* of the *Aktion Sühnezeichen* in 1983. In addition, a range of new memorial sites developed and new exhibitions were introduced at previously 'forgotten' concentration camps. Since most of these initiatives were bottom-up, the sites were often funded by local authorities, thus financial investment was limited. This approach changed with the fall of the Berlin Wall since unified Germany was suddenly confronted with the legacy of the GDR and the East German concentration camp memorials Ravensbrück, Buchenwald and Sachsenhausen.

After German unification, the politically influenced exhibitions at the East German concentration camp memorials were deemed inappropriate, and commissions were formed to re-design the memorial sites. This often led to heated debates, especially at Buchenwald and Sachsenhausen, where victims of the Soviet Special Camps (1945–1950) demanded to be heard. These commissions were dominated by historians from the former West Germany, thus implementing the 'Western model' of commemorating the Nazi past in the former East Germany. In fact, the West German focus in the 1980s on securing historic traces was transferred to the former East Germany. Comprehensive historical research commenced at the East German concentration camp memorials, which went hand in hand with a new landscape design with the aim of representing the sites in their 'truest' form. Since the new East German federal states were unable to meet the financial costs of implementing new exhibitions at the memorial sites, a decision was made to financially support the Eastern sites with 50 per cent from national funds. In addition, the national government funded the major memorials sites Topography of Terror and House of the Wannsee Conference, both in Berlin. This initial funding agreement in 1993 was thought to last for 10 years, yet Germany's responsibility for the memory of the two dictatorships and a search for a new German identity led to frequent debates about the future of the memorial sites in the German parliament.

The Enquete Commission for the *Überwindung der Folgen der SED-Diktatur im Prozess der deutschen Einheit* (overcoming of the consequences of the SED dictatorship in the process of German unity) was concerned about the future financial situation of the memorials sites that commemorate the GDR and therefore recommended in 1998 that financial support from the national government should continue as long as they fulfil certain criteria: located at historically authentic sites, have a scientific, museological and pedagogical concept, engage with victim associations and be of exceptional historical significance. In 1999, the overarching memorial concept (*Gedenkstättenkonzept*) was agreed upon by the SPD (Sozialdemokratische Partei Deutschlands – social democratic party of Germany) and the Greens coalition government under Gerhard Schröder. Nevertheless, given the political influence at the memorial sites in the GDR, the German government stressed the memorial sites' political independence, guaranteed by involving expert committees and victim associations. Hence, ironically, it was the collapse of the GDR and the responsibility for the East German *Mahn- und Gedenkstätten* that also secured the future of the sites in West Germany.

In 2008, the memorial concept was renewed, reiterating the original aims. The German government also emphasised its commitment to the so-called *Faulenbach Formula* which attempts to avoid a competition between victim groups and emphasises the differences between the SED-dictatorship and the Nazi regime. Dr Christina Weiss, the minister for culture and media in 2008, stressed that even the 'suspicion of equating different groups of victims is a relativisation, and anything that looks like a relativisation, like a relativisation of the National Socialist crimes against European Jews, can only damage Germany's reputation abroad' (cited in Garbe 2016). Thus, while Germany is committed to commemorating the victims of the GDR regime, its focus will remain on the Nazi regime. Hence, at this point, the memorial sites in the former West Germany (Bergen-Belsen, Neuengamme, Flossenbürg and Dachau) were added to the national scheme and were therefore eligible for national funding.

Although the original aim was to guarantee 50 per cent of national funding, the current reality differs. The *Stiftung Denkmal für die ermordeten Juden Europas* (Holocaust Memorial Berlin) receives 100 per cent and the *Gedenkstätte Deutscher Widerstand* (memorial site for the German resistance) 70 per cent of national funding. Consequently, representatives of the German government sit on the expert commissions that manage the sites. A special case is also the *Stiftung Sächsische Gedenkstätten* (Memorial foundation Saxony) where the entire foundation with its five memorial sites receives national funding rather than the individual sites. The funding for the former West German concentration camp memorials varies between 26 and 40 per cent. The German government also funds the *Bundesstiftung für die Aufarbeitung der SED-Diktatur* (Federal Foundation for the reappraisal of the SED dictatorship), which has no equivalent for the Nazi regime. Needless to say, this unequal funding leads to tensions between the memorial sites with those in the former West Germany lamenting the focus on the GDR and the decline of national funding in relative terms considering the rising costs due to inflation (Garbe 2016).

The West German concentration camp memorials never saw themselves as part of a tourism landscape. And although today's memorial managers recognise the increasingly international audience, they reject the notion of the concentration camp memorials becoming part of a tourism offer. Insa Eschebach (2016), Ravensbrück's former memorial manager, argues that while tourists might arrive with the expectation of wanting an 'authentic experience', the memorial sites should not pander to the wishes of the heritage or dark tourism industry. In fact, the staff teams should work against such expectations, especially during educational visits. The German historian, Bernd Faulenbach (2019), notes when reflecting on the work of the memorial sites since 1990, that Germany's sites increasingly have to battle with tourists' clichéd expectations of concentration camps, perpetrators and victims. He concludes that one should not reject (foreign) tourists at concentration camp memorials, but additional educational resources are needed to awaken empathy and historical knowledge in order to appreciate the site. Eschebach and Faulenbach, therefore, share the academic view of the 'shallow' tourist who requires additional support to understand the historical context. The irony of such viewpoints is that there are currently no offers for tourists at the memorial sites that underpinned the research described in this book. In fact, the only departments at memorial sites that cater to visitors are the educational departments whose sole focus is the school visitor. Yet this focus on cognitive outcomes ignores the importance of the emotional, spatial and social components of a visit to a memorial site (Pampel 2007). Natalie Bormann's (2018) study has shown that her American students were profoundly impacted by the eerie nature of Ravensbrück Concentration Camp memorial yet were somewhat disturbed by the touristy busyness of Dachau. In addition, Sharon MacDonald (2008) noticed during empirical research at the site of the Nazi Party Rally in Nuremberg that most visitors shared the opinion that 'being physically present is significantly different from learning about it from books' (169). Hence, for the visitor 'performing' a site is as important, if not more so, than any cognitive learning experience.

2.2 Gedenkstättenpädagogik

According to Germany's overarching memorial concept, *Gedenkstätten* are predominantly places of learning (*Lernorte*) and thus have to develop a 'wide reaching pedagogical offering' (Bundesregierung Online 1999). As such, they also adhere to the principles of the *Beutelsbacher Konsens*, a framework for political education that was developed in 1977 in West Germany. Although it is not a binding framework, it is now a piece of widely used guidance for political education in schools and the wider public sphere. Three core principles underpin the *Beutelsbacher Konsens*: firstly, a prohibition of emotionally overwhelming learners; secondly, the idea that anything that is controversial in politics and the economy needs to remain controversial; and finally that learners need to be enabled to develop their own political analysis from any given situation. In essence, the *Beutelsbacher Konsens* was designed to avoid politically influencing learners and to stimulate debates about controversial topics. For the memorial sites, the principle of not emotionally overwhelming visitors is of particular importance. Matthias Heyl (2016), Ravensbrück's educational manager, for instance, explains that teachers often demand a form of *Betroffenheitspädagogik* (engendering a form of sadness by using specific educational methods) during visits at the memorial sites that would lead to empathy with the victims, yet manipulating the students' emotions in this way would undermine the *Beutelsbacher Konsens* and is considered to be unethical. Particularly controversial in this regard is Hohenschönhausen's (Stasi prison memorial in Berlin) use of former inmates as tour guides, as their often emotionally intense tours do not enable visitors to form their own opinions. Hence, Hohenschönhausen was accused of violating the *Beutelsbacher Konsens* (Kowalczuk 2018). Verena Haug (2015), however, highlights that educational staff members at memorial sites often use the authority of the site to convey educational messages, thus constantly walking a fine line between using emotions for educational purposes while also being conscious not to emotionally overburden visitors.

The guiding principle of not overwhelming visitors influences the overarching management of the German memorial sites. It explains why exhibitions avoid emotionally engaging visitors by introducing interactive forms of engagement that are often found in exhibitions outside Germany, for example, ID cards of survivors at the Holocaust Museum in Washington, DC. Moreover, concepts that encourage the blurring of boundaries between visitors and former victims are rejected. In this regard Germany differs significantly from countries like the United Kingdom, which advocates the development of

'Holocaust ambassadors' who will pass on the knowledge of the Holocaust as witnesses of witnesses (Tollerton 2020).

With the introduction of Germany's memorial concept in 1999, the term *Gedenkstättenpädagogik* (education at memorial sites) was firmly established. Although a clear definition of the concept is pending, the memorial sites developed a common understanding of its aims (Haug 2015). Students should gain historical knowledge about the particular site, have an empathetic understanding of the victim and develop moral principles for political actions. Since the memorial sites often do not resemble former concentration camps, pedagogical staff also need to address the visual appearance of the site alongside codes for appropriate behaviour. At the centre of any educational session at the memorial sites stands the contextualisation of the Nazi regime, based on the perspective of perpetrators, victims and bystanders. By also increasingly focusing on bystanders and perpetrators, German memorial sites challenge the notion of the 'unexplainable atrocities of the Holocaust'. However, so far there is no common agreement at the concentration camp memorials on how closely the motivation of perpetrators should be taught. Moreover, German concentration camp memorials grapple with the difficulty of avoiding intense emotional reactions on the one hand and being unable to 'control' how visitors react on the other hand. Sachsenhausen's educational manager, therefore, suggests the establishment of safe spaces that allow students the sharing of difficult emotions (Schellenberg 2018). Such initiatives, however, assume that all educational visits are accompanied by pedagogical staff at the memorial sites when, in fact, some visits are conducted by schoolteachers themselves. Preliminary qualitative research with schoolteachers in Saxony, Saxony-Anhalt and Thuringia has revealed that teachers have an ambivalent attitude to the memorial sites' focus on cognitive learning (Münch 2019). The teachers expected an emotional reaction at the sites and even encouraged it at times. Yet for the memorial site's educational teams, a reflective understanding of the historical site is the most important outcome of a visit.

The literature on *Gedenkstättenpädagogik* neglects an important visitor to the memorial sites: the individual. Gisela Lehrke (cited in Behrens-Cobet 1998) already concluded in the 1980s that West Germany's educational activities were reduced to the school visitor ignoring the potential for adult learning. In fact, one can argue that since the 1980s not much progress was made since the debates in the current literature centre around pedagogical strategies of school visitors. While educational staff can 'control' pupils' behaviour during school visits, they are unable to control the emotional responses of individual visitors. The belief is that the focus of cognitive learning outcomes in the exhibition design will elicit strong emotional reactions. Brink (1998), for instance, explains that in the early years of exhibition design at the memorial sites, it was common practice to use emotionally charged images, often oversized, with the aim of pointing the moral finger at the German population that allowed such atrocities to take place. Yet the generational shift and a more diverse society demanded a different approach. Today's exhibitions are characterised by objectivity and a scholarly representation of the past that values learning above enjoyment (Yair 2014). In this context, Gad Yair's use of the word 'enjoyment' refers to the emotional connection visitors often make in museum spaces, which are discouraged at Germany's memorial sites. In fact, with his Israeli background, he considers Germany's factual approach to exhibition design to be offensive. Yair's reaction reveals an unease with which Germany's memorial managers grappled for some time: how does one hold onto the notion of an emotional detachment while increasingly accommodating visitors who have their own traumatic memories and who very often draw a comparison to present-day traumas, therefore challenging Germany's notion of the singularity of the Holocaust. The answer to this dilemma is pending. Elke Gryglewski (2010), the former acting director of the House of the Wannsee Conference, argues that one needs to distinguish between a political motivation for drawing comparisons and an individual perspective, with the latter potentially not requiring a corrective answer. Moreover, educational staff who deliberately use comparisons to present-day events should interrogate their own motivations for doing so and consider the purpose of it in light of educational aims. In essence, Germany's memorial sites actively discourage visitors from reflecting on contemporary challenging political situations.

This issue becomes even more pertinent when considering the increasingly diverse German society. Esra Özyürek (2018) points out that from the 2000s onwards newspapers began to run stories about Muslim communities not acknowledging the Holocaust appropriately. They are accused of either ignoring the Holocaust or not having learned the correct lessons. Yet Germans with a Turkish or Arab background approach Holocaust history from their own perspective of, for instance, facing daily racism, or having had their own traumatic memories of violence and war. Thus, their perception of the Holocaust is from the point of view of a victim, rather than the German expectation of remorse and responsibility. And although German pedagogical staff recognise that not every German has roots within the Third Reich, Germany's memory culture lacks the flexibility to include communities who respond differently to the Holocaust.

When reviewing the academic literature on *Gedenkstättenpädagogik*, it also becomes apparent that the concept is predominantly discussed in relation to the Nazi regime. A discussion about pedagogical strategies at the former Stasi prison memorials is absent. Whilst Hohenschönhausen is criticised for its use of former prisoners as tour guides and/or emotional strategies such as locking visitors into cells, there are no academic debates about the appropriate strategies of education at the memorial sites that commemorate the GDR past. Klaus-Dieter Kaiser (2016) argues that there is a common consensus now in Germany that the Nazi regime was an unlawful state and that denial would be a criminal offense; yet for the GDR such common ground is a long way off. He is, therefore, implying that questioning the GDR dictatorship should become unlawful. Yet the GDR's legacy is contested precisely because it does not have the characteristics of a regime that committed a large-scale genocide, making comparisons with the Third Reich inappropriate.

2.3 The GDR memory debates

Carola Rudnick (2014) argues that dealing with the legacy of the GDR was swift and had already been initiated by the last GDR government under Hans Modrow. Initially, this form of reckoning with the GDR focused on justice processes for victims of the SED regime (e.g. political prisoners). Although roughly 62,000 cases were investigated, only a small number of them were actually brought to court (Clarke and Wölfel 2011). Thus, the focus in the early years was on the Stasi (Ministry of State Security) and its oppressive apparatus, enabled by the easy access to individual Stasi files. In addition, the GDR's archive was immediately transferred to the West German federal state archives and opened without the usual 30-year delay. This led to a 'history boom' with no other period in history so quickly researched than East Germany. Alongside the legal prosecutions, Germany's strategy was to remove Communist symbols: street names were changed, statues were taken away, the Berlin Wall was demolished and official GDR remembrance days were discontinued (Rudnick 2014). During this phase, there was little appetite for considering future commemorative practices.

With the rise of *Ostalgie* in the mid-1990s, a play on words that describe the yearning for the GDR, the attitude changed. Conservative politicians were concerned that romanticising the GDR would lead to a failure of democratising the East (Clarke and Wölfel 2011; Rudnick 2014). As a consequence, the first Enquete Commission enquiry into the history of the GDR was launched. The nature of the enquiry was directed at the repressive regime, opposition and resistance; thus, politically the aim was to delegitimise the GDR as an *Unrechtsstaat* (a state without law). The societal areas of culture, welfare and economy were regarded as secondary. Whilst many consider such state involvement inappropriate, Andrew Beattie (2011) states that the organisation was democratised by allowing members from all parties, including East Germans, to take part. Nevertheless, two opposing viewpoints stood against each other: the anti-communist, anti-fascist and totalitarian theory of the CDU/CSU and the liberal, communist-critical theory of the PDS (Partei der sozialistischen Demokratie – party of democratic socialism) and the SPD (Rudnick 2014). As a consequence, the discourse of *Diktaturgedächtnis* (the GDR as a dictatorship) dominated, and subsequently the first Enquete Commission, which operated between 1992 and 1994, confirmed its pre-determined outcome: the GDR was indeed a dictatorship. In remembering the GDR, however, the so-called *Faulenbach-Formula* was to be applied; it suggests that 'Nazi crimes must not be relativized by dealing with the crimes of Stalinism, but the memory of the crimes under the Soviet occupation and the SED rule should not be trivialized with reference to the Nazi regime' (Faulenbach cited in Pearce 2011, 177). Whilst this approach is commendable, in reality it is difficult to achieve, as I will later discuss in relation to the Stasi prison Bautzen II in Chapter 4.

The second Enquete Commission was launched in 1995 and dealt with the themes that had been previously neglected: education and science in the

GDR, social, economic and environmental policies, and the role of a divided Germany within Europe. The final report culminated in the recommendation of an overarching memorial concept. It was a significant step for Germany as it linked the memorial sites to democratic development and overall responsibility for a united Germany (Pearce 2011). Thus, the national memorial concept paved the way for an 'institutionalised memory'. In memorialising the GDR, the focus was initially on Hohenschönhausen, and yet it was the one site that was not known in the GDR. Only after the second Enquete Commission were Bautzen II, Torgau, the German-German Museum Mödlareuth, the Inner German Border Museum Marienborn and Haus I Normannenstraße added. Sites such as Chemnitz-Kaßberg or Hoheneck which are deeply anchored in East German memory were excluded.

In 2004, societal shifts led to the formation of a new commission with the aim of reviewing the memory landscape of the former GDR, the Sabrow Commission. Unlike the *Bundestag* Enquete Commissions, experts were appointed by the federal commissioner or co-opted by the commission itself. Unfortunately, debates were held behind closed doors. Consequently, the Sabrow Commission had little political legitimacy and its final report, published in 2006, was heavily criticised. Martin Sabrow distinguished between three ideal memory types: the dictatorship and the peaceful revolution of 1989, daily life and GDR citizens' arrangement with the regime, and progress in the GDR from the Nazi past to the fall of the Berlin Wall (Meyen and Pfaff-Rüdiger 2014). In the dictatorship model, the narrative is focused on the victim, perpetrator and resistance, with no room for the cultural life in the GDR. It goes without saying that this model is the preferred discourse in the German media. The second model considers self-assertion under tough conditions and is a memory type that is very vivid in East Germany today. The third model, the GDR and its achievements from the Second World War to reunification, is very popular amongst the former GDR elites. Although Sabrow's models were not widely accepted, the renewal of Germany's memorial concept in 2008 incorporated some of Sabrow's suggestions. The GDR memorial landscape would be divided into four distinct categories: division and border, surveillance and oppression, society and everyday life, and resistance and oppression (Wüstenberg 2017). One could argue that today's GDR memory landscape has somewhat solidified, yet debates about the GDR legacy erupt regularly.

Anna Kaminsky (2020) explains that with the fall of the Berlin Wall two cultures of memory collided: the 'Western' memory with a focus on the Nazi regime and the Eastern European memory with a stronger focus on the 'Gulag' system. The Soviet Union's double role as a liberator of Nazi Germany and as a leader of an inhumane regime has dominated the German discourse ever since unification. Moreover, victims who suffered in the Soviet Special Camps or in the GDR's Stasi prisons regularly demand greater recognition. Thus, the focus remains on the memory of the Stasi and the subsequent treatment of the Stasi files. Ilko-Sascha Kowalczuk (2021) further emphasised that the hatred against the Stasi was omnipresent in the GDR; so unsurprisingly with the fall of the

Berlin Wall this hatred was given an outlet. The difficulty is that it absolved the SED-apparatus of any responsibility, since without them the Ministry of State Security could not have operated. In the meantime, however, the generation who suffered under the particularly harsh conditions of Walter Ulbricht's GDR (1949-1971) diminishes, leaving behind Honecker's (1971−1989) children with very different memories. This generation's GDR memory is clouded by the economic and social hardship following its collapse, a memory that is barely heard in the official discourse.

Michael Meyen and Senta Pfaff-Rüdiger (2014) examined GDR memory under Assmann's concept of communicative memory by using focus-group research. The findings proved that most Germans were aware of the official dictatorship discourse, but conclusions differed. West Germans felt no need for further explanations, whilst East Germans avoided the narrative as they did not want to be represented as coming from a lawless state under the rule of the Stasi. These findings correspond with Wolf Wagner's (2015) analysis of the East German state of mind. Wagner (*ibid*) points out that the absorption of East Germany into the Federal Republic of Germany has created a culture shock. Almost overnight, the life many East Germans had known vanished and they found themselves in a completely different culture without having moved an inch. In addition, the GDP (Gross Domestic Product) dropped in East Germany by 46 per cent, which is comparable economically to a lost war. This traumatic experience which went hand in hand with the social decline in East Germany has consequences to this day. For instance, East Germans consistently refer to themselves as second-rate citizens, a perception that has changed very little.

Saxony's former minister for Integration Petra Köpping (2018), therefore, argues that remembering the GDR is obfuscated. The focus on the peaceful revolution in 1989 completely ignores the pain of the economic difficulties in East Germany following reunification. Moreover, the focus of the GDR as an *Unrechtsstaat* has caused damage as many East Germans do not recognise this narrative (Hensel 2018). Hence, Kowalczuk (2018) argues that the 'old' historians need to retire, allowing a new generation with less emotional baggage to come forward. Thus, 30 years after German reunification, memory of the GDR is not a 'cold' one as previously predicted (Maier 2002). In fact, several organisations formed by East Germans who were born either shortly before or after the fall of the Berlin Wall argue for a differentiated image of GDR history (e.g. Third Generation East) and demand an enquiry into the work of the *Treuhand*, the organisation that privatised the GDR assets, which resulted in a highly unequal wealth transfer to the former West Germany.

In light of the widening gap between the former East and West Germany, a commission consisting of academics and politicians was launched as part of the 30th anniversary of German reunification with the aim of critically reflecting on the transformation process in East Germany (Bundesministerium des Innern, für Bau und Heimat 2020). They recommended the creation of an international centre for transformation, which would shine a light on the challenges East Germans had to face and would engage in research that addresses contemporary transformation processes, for example, climate change. The commission also suggests, among many other recommendations, providing better recognition and psychological care for the victims of the GDR. Indeed, with agreed funding for two new GDR memorial sites, Chemnitz-Kaßberg and Hoheneck due to open in 2022, neglected parts of GDR history will be addressed. Chemnitz-Kaßberg will engage with the issue of *Freikauf* (the 'purchasing' of GDR political prisoners by West Germany), while Hoheneck, the former GDR's women's prison, will confront the female experience of political imprisonment. On the other hand, with increasing distance, memories of the GDR become more differentiated. In quick succession, two books were published in 2020 that examine the GDR's art and culture and thus challenging the dominant dictatorship discourse (Decker 2020; Martin 2020). In essence, historical memory of the GDR remains a contested field.

2.4 Bibliography

Arnold-de Simine, Silke. 2012. 'Memory Museum and Museum Text: Intermediality in Daniel Libeskind's Jewish Museum and W.G. Sebald's Austerlitz'. *Theory, Culture & Society* 29 (1): 14–35.

Beattie, Andrew. 2011. 'The Politics of Remembering the GDR: Official and State-Mandated Memory since 1990'. In *Remembering the German Democratic Republic: Divided Memory in a United Germany*, edited by David Clarke and Ute Wölfel, 23–34. Basingstoke: Palgrave MacMillan.

Behrens-Cobet, Heidi. 1998. 'Erwachsene in Gedenkstätten – Randständige Addressaten. Zur Einführung'. In *Bilden und Gedenken. Erwachsenenbildung in Gedenkstätten und an Gedächtnisorten*, edited by Heidi Behrens-Cobet, 7–22. Essen: Klartext-Verlag.

Bormann, Natalie. 2018. *The Ethics of Teaching at Sites of Violence and Trauma – Student Encounters with the Holocaust*. New York: Palgrave Macmillan.

Brink, Cornelia. 1998. *Ikonen der Vernichtung: Zum öffentlichen Gebrauch von Fotografien aus nationalsozialistischen Konzentrationslagern nach 1945*. Berlin: De Gruyter.

Bundesministerium des Innern, für Bau und Heimat. 2020. *Abschlussbericht der Kommission "30 Jahre Friedliche Revolution und Deutsche Einheit"*. Berlin: Bundesministerium des Innern, für Bau und Heimat. https://www.bmi.bund.de/SharedDocs/downloads/DE/veroeffentlichungen/2020/abschlussbericht-kommission-30-jahre.html.

Bundesregierung Online. 1999. 'Konzeption der künftigen Gedenkstättenförderung des Bundes'. 27 July. http://dip21.bundestag.de/dip21/btd/14/015/1401569.pdf.

Clarke, David, and Ute Wölfel. 2011. 'Remembering the German Democratic Republic in a United Germany'. In *Remembering the German Democratic Republic: Divided Memory in United Germany*, 3–22. Chippenham: Palgrave Macmillan.

Decker, Gunnar. 2020. *Zwischen den Zeiten: Die späten Jahre der DDR*. Berlin: Aufbau Verlag.

Eschebach, Insa. 2016. 'Einführung'. In *Von Mahnstätten über zeithistorische Museen zu Orten des Massentourismus: Gedenkstätten an Orten von NS-Verbrechen in Polen und Deutschland*, edited by Enrico Heitzer, Günter Morsch, Robert Traba, and Katarzyna Woniak, 25–27. Berlin: Metropol-Verlag.

Faulenbach, Bernd. 2019. 'Eine neue Erinnerungskultur? – Entwicklungslinien und Probleme der Gedenkstätten seit der Epochenwende 1989/90'. *Sachsenhausen Lectures* 3 (1): 1–44.

Garbe, Detlef. 2016. 'Die Gedenkstättenförderung des Bundes: Förderinstrument im geschichtspolitischen Spannungsfeld'. *Gedenkstättenrundbrief* 182: 3–17.

Gryglewski, Elke. 2010. 'Teaching about the Holocaust in Multicultural Societies: Appreciating the Learner'. *Intercultural Education* 21 (1): 41–49.

Haug, Verena. 2015. *Am "authentischen" Ort: Paradoxien der Gedenkstättenpädagogik.* Berlin: Metropol-Verlag.

Hensel, Jana. 2018. '"Kulturgeschichte der DDR": "Erstaunlich, wie lange dieses Land existiert hat"'. *Die Zeit*, 22 December 2018. https://www.zeit.de/2018/53/kulturgeschichte-der-ddr-gerd-dietrich-historiker-forschung.

Heyl, Matthias. 2016. 'Mit Überwältigendem überwältigen? Emotionalität und Kontroversität in der historisch-politischen Bildung. Ein Plädoyer für die Schärfung des Profils historischer Bildung'. In *Politische Bildung auf schwierigem Terrain: Rechtsextremismus, Gedenkstättenarbeit, DDR-Aufarbeitung und der Beutelsbacher Konsens*, edited by Jochen Schmidt, Steffen Schoon, and Landeszentrale für Politische Bildung Mecklenburg-Vorpommern, 37–55. Schwerin: Landeszentrale für Politische Bildung Mecklenburg-Vorpommern.

Kaiser, Klaus-Dieter. 2016. 'Was bedeutet der Beutelsbacher Konsens als didaktische Grundlage im Blick auf die Vermittlung von DDR-Geschichte'. In *Politische Bildung auf schwierigem Terrain: Rechtsextremismus, Gedenkstättenarbeit, DDR-Aufarbeitung und der Beutelsbacher Konsens*, edited by Jochen Schmidt, Steffen Schoon, and Landeszentrale für Politische Bildung Mecklenburg-Vorpommern, 37–58. Schwerin: Landeszentrale für Politische Bildung Mecklenburg-Vorpommern.

Kaminsky, Anna. 2020. 'In der Mitte der Gesellschaft angekommen? Die Auseinandersetzung mit der kommunistischen Diktatur in der SBZ und DDR im Vereinigten Deutschland'. In *Erinnerungs- und Gedenkorte im sächsischen Dreiländereck Polen-Tschechien-Deutschland*, 65–91. Dresden: Sächsische Landeszentrale für politische Bildung/Umweltbibliothek Großhennersdorf e.V.

Knoch, Habbo. 2018. 'Gedenkstätten'. *ZZF: Centre for Contemporary History*. https://doi.org/10.14765/zzf.dok.2.1221.v1.

———. 2020. *Geschichte in Gedenkstätten: Theorie – Praxis – Berufsfelder*. Tübingen: UTB GmbH.

Köpping, Petra. 2018. *'Integriert doch erst mal uns!' – Eine Streitschrift für den Osten*. Berlin: Ch. Links Verlag.

Kowalczuk, Ilko-Sascha. 2018. 'Und was hast du bis 1989 getan?'. *Süddeutsche Zeitung*, 23 October. www.sueddeutsche.de/kultur/ddr-geschichte-aufarbeitung-1.4179958?reduced=true.

———. 2021. 'Das Nicht-vertrauen-Können belastet Ostdeutschland bis heute'. *Berliner Zeitung*, 1 February. www.berliner-zeitung.de/zeitenwende/das-nicht-vertrauen-koennen-belastet-ostdeutschland-bis-heute-li.135563.

Leo, Annette. 1998. 'Geschichtsbewußtsein "herstellen" – Ein Rückblick auf Gedenkstättenarbeit in der DDR'. In *Bilden und Gedenken. Erwachsenenbildung in Gedenkstätten und an Gedächtnisorten*, edited by Behrens-Cobet Heidi, 35–50. Essen: Klartext-Verlag.

Macdonald, Sharon. 2008. *Difficult Heritage: Negotiating the Nazi Past in Nuremberg and Beyond*. Abingdon: Routledge.

Maier, Charles S. 2002. 'Hot Memory . . . Cold Memory: On the Political Half-Life of Fascist and Communist Memory'. *IWM* (blog). www.iwm.at/transit/transit-online/hot-memory-cold-memory-on-the-political-half-life-of-fascist-and-communist-memory/.

Martin, Marko. 2020. *Die verdrängte Zeit: Vom Verschwinden und Entdecken der Kultur des Osten*. Stuttgart: Tropen.

Meyen, Michael, and Senta Pfaff-Rüdiger. 2014. 'Mass Media and Memory: The Communist GDR in Today's Communicative Memory'. *Medijske Studije* 5 (9): 3–17.

Münch, Daniel. 2019. 'Gedenkstättenbesuche als emotionales Erlebnis. Welche Rolle weisen Geschichtslehrkräfte den Emotionen ihrer Schülerinnen und Schüler zu?' In *Holocaust Education Revisited. Wahrnehmung und Vermittlung – Fiktion und Fakten – Medialität und Digitalität*, edited by Anja Ballis and Markus Gloe, 87–108. Wiesbaden: Springer VS.

Niven, Bill. 2007. *The Buchenwald Child: Truth, Fiction, and Propaganda*. Rochester, NY: Camden House.

———. 2009. 'Remembering Nazi Anti-Semitism in the GDR'. In *Memorialization in Germany since 1945*, edited by Bill Niven and Chloe Paver, 205–212. London: Palgrave Macmillan.

Özyürek, Esra. 2018. 'Rethinking Empathy: Emotions Triggered by the Holocaust among the Muslim-Minority in Germany'. *Anthropological Theory* 18 (4): 456–477.

Pampel, Bert. 2007. *'Mit eigenen Augen sehen, wozu der Mensch fähig ist': Zur Wirkung von Gedenkstätten auf ihre Besucher*. Frankfurt am Main: Campus Verlag.

Pearce, Caroline. 2011. 'An Unequal Balance? Memorialising Germany's "Double Past" since 1990'. In *The GDR Remembered: Representations of the East German State since 1989*, edited by Nick Hodgin and Caroline Pearce, 172–201. Rochester: Camden House.

Piltz, Georg. 1985. *Kunstführer durch die DDR*. Berlin, Jena: Urania Verlag.

Rudnick, Carola S. 2014. *Die andere Hälfte der Erinnerung: Die DDR in der deutschen Geschichtspolitik nach 1989*. Bielefeld: Transcript Verlag.

Schellenberg, Martin. 2018. 'Beyond Learning the Facts: Teaching Commemoration as an Educational Task in German Memorial Sites for the Victims of National Socialist Crimes'. In *Remembering the Holocaust in Educational Settings*, edited by Andy Pearce, translated by Mathew Turner, 122–139. London; New York: Routledge.

Tollerton, David. 2020. *Holocaust Memory and Britain's Religious-Secular Landscape: Politics, Sacrality, and Diversity*. New York: Routledge.

Wagner, Wolf. 2015. 'Unification by Absorption or by Incrementalism (Sunshine Policy)?: A Comparative Enquiry 25 Years after German Reunification'. *Development and Society* 44 (1): 167–189.

Williams, Paul. 2008. *Memorial Museums: The Global Rush to Commemorate Atrocities*. New York: Bloomsbury Academic.

Wüstenberg, Jenny. 2017. *Civil Society and Memory in Postwar Germany*. Cambridge: Cambridge.

Yair, Gad. 2014. 'Neutrality, Objectivity, and Dissociation: Cultural Trauma and Educational Messages in German Holocaust Memorial Sites and Documentation Centers'. *Holocaust and Genocide Studies* 28 (3): 482–509.

3 The memorial sites of Flossenbürg, Ravensbrück, House of the Wannsee Conference and Bautzen II

The chosen case studies for this research project are representative of the intense memorialisation processes Germany has undergone since 1945 to commemorate, and face up to, the Nazi past and more recently the GDR. In this chapter, I will demonstrate how German memory politics have influenced the development of the memorial sites. Flossenbürg and the House of the Wannsee Conference, both located in the former West Germany, were sites that were largely forgotten. At Flossenbürg, a deliberate destruction programme erased almost all traces of the former concentration camp: out of sight, out of mind. Although the Wannsee Villa, the location of the Wannsee Conference, escaped the same fate, it vanished from people's memory. Initially converted into a conference centre for German social democrats, it was later used as a recreational home for children from deprived backgrounds. Attempts to convert the site into a memorial in the 1960s were met with overall disapproval; how could the children's laughter be destroyed by establishing a Nazi memorial was the sentiment expressed at the time. Ravensbrück Concentration Camp, located in the former East Germany, was also subject to transformation, albeit very differently. Liberated by the Red Army, large sections of the site were changed into a Soviet army base. This restricted access to most of the former camp. But a small area, comprising the former solitary confinement block and the crematorium, formed part of the first memorial opened in 1959. Bautzen II, the Stasi prison memorial, wrestles with the memory of the most recent past: the GDR dictatorship. Established as a memorial very quickly after reunification, its challenges lie in the integration of different time periods, each of them is marked by its own significant atrocities: the Nazi prison, the Soviet Special Camp and the Stasi prison. This causes challenges when designing exhibitions as, for instance, the history of the Soviet Special Camp is barely known. I will therefore also explain in this chapter how the exhibitions were designed and what impact this will have on the overall visitor experience.

3.1 The development of Flossenbürg Concentration Camp memorial

Flossenbürg is situated in North Bavaria, a region called *Oberpfälzer Wald*, very close to the border of the former Czechoslovakia. In the 1930s, the region

DOI: 10.4324/9781003126836-3

was underdeveloped and, together with the cold Northern wind, gave the area the nickname 'the German Siberia'. Flossenbürg, however, was an exception. It possessed large areas of granite and since the 19th century various quarries had opened onsite (Heigl 1989). Hitler's extensive building programme in the 1930s increased the demand for raw materials, and Flossenbürg with its extensive granite quarry became the ideal location for a forced labour camp, opened 21 April 1938. Alongside the concentration camp, the SS-owned company *Deutsche Erd- und Steinwerke GmbH* was launched, the organisation that oversaw the quarrying and the supply of forced labour.

The first Flossenbürg prisoners arrived from Dachau and were primarily responsible for the erection of the camp infrastructure: further barracks, administration and laundry buildings (Benz et al. 2007). In these early days, over half of them had a previous criminal background; they had been convicted prior to the establishment of the concentration camp system and were not released at the end of their sentence. Therefore, Flossenbürg inherited the title 'camp for asocials and criminals'. This influenced the public opinion that the prisoners at the camp deserved their imprisonment, a perception which can still be found today amongst the local community.

In 1940, Flossenbürg was changed into a category II camp (equivalent to Buchenwald, Neuengamme and Auschwitz II) for prisoners who could not be 'improved' (*ibid*). Subsequently, the first Jewish prisoners arrived, and in January 1941, the first Polish prisoners were transferred from Auschwitz to Flossenbürg. In addition, from 1941 onwards, a prisoner-of-war camp was established for Soviet soldiers, who were often executed. By 1944, 8,000 prisoners were at Flossenbürg; the largest group consisted of Polish prisoners followed by Soviet prisoners of war, French, Belgian and Dutch prisoners. As the demand for building materials had declined, most prisoners were now working at the *Messerschmidt GmbH* which produced ME-109 fighter planes.

As the Allied Forces advanced into Germany in March 1945, the evacuation of the camps commenced (Heigl 1989). At Flossenbürg, initially all Jewish prisoners were evacuated, 'loaded' on to trains and transported south. On the journey, the train was attacked by American fighter planes and most prisoners died. More evacuations followed, for example 'special' prisoners (publicly known hostages) were transported to Dachau. Finally, on 22 April 1945, the American army reached Flossenbürg. The village itself had already surrendered, and at the camp the American army was greeted with the banner in English 'Prisoners. Happy End. Welcome!' as SS-staff had already left the site. American soldiers were confronted with the horrors of the concentration camp, though the initial American reports only describe the facilities of the site in a clinical way. After the liberation, the American army delivered medical care to the camp and established a field hospital as the remaining prisoners (approximately 1,600) suffered from typhus, tuberculosis and other infectious diseases. However, despite the care, during the first week 30 prisoners died every day. By the end of May 1945, epidemics were under control and prisoners were subsequently repatriated. Flossenbürg concentration camp was closed.

Flossenbürg played a significant role within the concentration camp system. It was the centre of an extensive network of 90 subcamps reaching as far as Saxony and the Czech Republic. In total, 90,000 prisoners were at Flossenbürg and its subcamps, of which approximately 30,000 did not survive. Flossenbürg was also the model for other subsequent forced labour camps; indeed, it was the first site where the slogan *Arbeit macht frei* was thought to have been displayed (Stier 2015). Yet after its liberation, it vanished from people's memory and became known as the 'forgotten concentration camp' (Skriebeleit 2011).

3.1.1 The development of the memorial site

After the camp's liberation, a committee had formed (executive committee for erecting the monument and chapel at Flossenbürg Concentration Camp), consisting of people living in the UNRRA (United Nations Relief and Rehabilitation Administration) camp and officials from local authorities (Skriebeleit 2009). Polish Catholics dominated the committee to the exclusion of Jewish members who had little influence on the development of the first memorial. It is also worth noting that the committee consisted of people who had experienced imprisonment in concentration camps, but not in Flossenbürg.

Eventually, two different commemorative practices were developed: a memorial at the cemetery which had already been established by the Americans in the centre of the village, and a Catholic chapel in the former area of the concentration camp (Skriebeleit 2009). After a fundraising campaign by the committee, a monument was erected at the cemetery (*Ehrenfriedhof*) in the village. The memorial includes the inscription 'In Memoriam Consortes' and lists the number of victims and their nationalities. The planning department of the Bavarian Interior Ministry criticised the structure of the memorial as it destroyed the surrounding landscape, bearing in mind its location in the centre of the village (Heigl 1989). Indeed, the concentration camp was located outside the village; hence, the local community could avoid it if they wished. With the establishment of the memorial, however, the concentration camp 'travelled' to the heart of the village, confronting the local community.

At the former concentration camp, the local architect, Christian Lindhardt, was tasked with the design of a new chapel, which was supposed to commemorate the different nations who had suffered at the hands of the Nazis. He suggested incorporating the chapel into the valley together with the former crematorium, calling it collectively the valley of death (Skriebeleit 2009). The path across the valley would resemble the way of the cross: a descent into hell signified by the crematorium and an ascent to salvation symbolised by the chapel (Skriebeleit 2016). Moreover, the memorial was not to disturb the beauty of the surrounding woodlands (Skriebeleit 2009). Metaphorically, the valley served two purposes: as a container it would avoid the spilling over of traumatic memories to the wider landscape, and as a boundary between the dead and the living, so life could go on undisturbed. The chapel was named 'Jesus in prison' by the Bishop in Regensburg, highlighting the Polish Catholic influence and the shift away from the victim to Christian symbolisms. With the opening of the chapel, Flossenbürg was the first European memorial at a former concentration camp. Yet it was completely disassociated from the actual reality of the camp.

In the coming years, the remainder of the former concentration camp was transformed into a housing and industrial estate. The barracks, re-used to accommodate refugees from the Eastern territories, gave way to a modern housing estate. Although it provided much-needed homes for the refugees, it also excluded 'the newcomers' from daily life in the village since the former concentration camp was located on the outskirts. The *Kommandantur* was converted into social housing, while other buildings provided employment opportunities as industrial units. Designed to blend in with the landscape, the 'attractive and luxurious' SS houses were sold as private residencies. Since most prisoners were from Eastern Europe and therefore behind the Iron Curtain, an active survivors' network did not develop in Flossenbürg, and the removal of historical fabric was not met with any resistance. With the increasing popularity of the theologian Dietrich Bonhoeffer, who was hanged in Flossenbürg several days before liberation, the German Lutheran church suddenly demanded a commemorative exhibition. Nobody was therefore overly concerned when this new exhibition finally opened in 1969, focusing exclusively on Dietrich Bonhoeffer and Wilhelm Canaris while ignoring other victim groups. Bonhoeffer's popularity led to increased tourism to Flossenbürg in the 1970s, resulting in a changed attitude amongst the local population. Whilst the memorial had previously been viewed as an eyesore, it was now becoming an important part of Flossenbürg's tourism landscape. It was, of course, easier to commemorate the heroes of the Nazi resistance than to acknowledge the atrocities committed on 'ordinary' prisoners.

The fall of the Berlin Wall created a significant shift in German memory politics, which also impacted on Flossenbürg's development. For the 50th anniversary of the end of the Second World War, a large number of survivors from the former Eastern bloc travelled to Flossenbürg to take part in the commemorations. The event was widely reported in national and local media, resulting in an increased public interest. When in 1999 the former concentration camp buildings, which had been used as industrial units, were returned to the Bavarian government, the opportunity arose to re-establish the camp outline and provide a permanent memorial onsite.

The building work commenced in 2004, demolishing old factory buildings and returning the site as closely as possible to its 1945 appearance. The *Appellplatz* was reinstated by removing buildings and trees, creating a barren landscape. The location of the former barracks was indicated by white lines and the camp boundary was reinstated by using white concrete fence posts. Since 2004, the strategy has therefore been to make evident the traces of mass murder. In fact, the landscape has turned to a previous moment in time, precisely to the date of the liberation in April 1945. As such, Flossenbürg is a 'frozen' time capsule, although nature had different ideas. It reclaimed the area and transformed it into a picturesque valley where the aesthetics of the landscape stands in stark contrast to the atrocities exhibited. The natural progress, however, was interrupted by removing natural features to reinstate camp features, enshrining the trauma forever into the landscape. Nevertheless, for the French Victim Association even this development was not acceptable: 'We demand calibrated gravel (like Buchenwald) instead of grass and the restoration of original fence posts erected to the right of the gate with various barbed wires' (Association des Déportés et Familles de Disparus du Camp de Flossenbürg cited in Skriebeleit 2016, 48). The disagreement between the management team at Flossenbürg and the French Victim Association led eventually to the president's resignation, due to the sheer outrage about Flossenbürg's inability to compromise (*ibid*). This example shows that the restoration of a memorial landscape is not a straightforward process. Indeed, the victim association seemed to have preferred a 'lifeless' landscape as displayed at the former GDR memorial sites. Furthermore, the housing development, still intact, is now considered to be highly inappropriate, with some people suggesting compulsory purchase with a long-term plan to demolish all houses. Considering that most residents in the houses are refugees or children of refugees, who have already lost their homes during the expulsion process in 1945, uprooting them again would be very traumatic.

Thus, Flossenbürg is a very complex memorial landscape which a tourist to the memorial site will have to disentangle. Firstly, two very different forms of memorialisation are merging: the early commemorative attempt with a focus on Christian redemption, and today's concept with the aim of making evident the traces of past suffering. Moreover, the housing development on the foundations of the barracks features names such as 'Sudetenstrasse' and 'Schlesierweg', emphasising the German pain. On the other hand, residents also possess positive memories of playing in the barracks as children, finally calling Flossenbürg 'home' again (Möller 2019). Thus, Flossenbürg's memorial landscape can be best described as a palimpsest with several layers overlapping each other.

3.1.2 The current memorial and its exhibitions

The visitor enters the site via the main *Kommandantur* building that houses the staff offices for the site. Since there is no visitor centre and/or reception there is little sense of an 'arrival', and no possibility to 'prepare' visitors for the site. The visitor is then immediately faced with the new housing development before s/he continues to walk across the former *Appellplatz* to the former prison kitchen and laundry, the only buildings left in this area. Visitors are able to explore freely the memorial complex, including the valley of death with the crematorium, the ash pyramid (mass grave) and the chapel, the former *Sonderblock* area (special block, the camp's brothel) and the former camp prison (*Arrestbau*) where Bonhoeffer was hanged.

The former laundry houses the exhibition 'Flossenbürg Concentration Camp 1938–1945'. It details the historical development of the concentration camp and its satellite camps using original objects, historical documents and eyewitness accounts. Flossenbürg's exhibition differs from Germany's usual factual approach to exhibition design. An interactive model at the beginning of the exhibitions shows the development of the camp, while extensive video footage of the death marches confronts the visitor with some harrowing images before s/he leaves the exhibition. In fact, Flossenbürg does not shy away from using controversial images, such as prisoners hanging on a Christmas tree. Therefore, Flossenbürg somewhat defies the principles of the *Beutelsbacher Konsens* of not emotionally overburdening visitors. Jörg Skriebeleit (2016, personal communication), the current memorial manager, admits that Flossenbürg was criticised for this approach, but by showing the challenging material of the death marches, for instance, he wanted to highlight the German bystanders. In the basement of the laundry individual stories represent the different victim groups, reiterated by the film *We have survived – The others had to stay* which is shown in the cinema. The former prisoners' shower room is retained in its original format, and only a small interpretation panel explains the significance of this room. Here the memorial team relies on historical authenticity to convey the room's significance; it does not pass visitors by unnoticed as I will show later.

The second permanent exhibition is based in the former kitchen area and deals with the consequences of the concentration camp after 1945, in particular what remained of the victims and of the perpetrators. In this regard, the exhibition is unique in that the memorial site takes a critical stance towards its own development, but also demonstrates Germany's way of coming to terms with the past. It acknowledges the era of silence, the purposeful removal of historical evidence to avoid the stigma, as well as envisioning the challenges that Germany faced, and still faces, when confronting the legacy of the Nazi past. Nevertheless, it is an academically challenging exhibition as the content requires a basic knowledge of memory politics in the two German states, with which non-German visitors might not be familiar.

Due to the removal of buildings and the re-use of the camp for industrial purposes after 1945, there is little evidence of the conditions of the concentration camp during the Nazi period. As such, the current memorial does not reflect the typical atmosphere a visitor might expect. In addition, both exhibitions were created with modern design techniques and do not resemble the historical conditions of the laundry, prisoners' bath and the kitchen. In fact, one could say that the process of visual improvement in the 1960s and 1970s was successful as it created a sanitised version of a concentration camp.

Flossenbürg is a living reminder of the memory politics in West Germany between 1945 and 1995. This commenced with a period of erasure with the council identifying itself as the victim, followed by a period of selective memory focusing on the Nazi resistance. Flossenbürg also highlights the enormous influence of victim associations elsewhere in establishing memorial sites in Germany. The lack of active involvement at Flossenbürg paved the way for a systematic removal of historical evidence without major complaints. And whilst the current memorial site attempts to accurately present the former camp and its history, even winning the Museum of the Year Award 2014 for its brave exhibition on Flossenbürg after 1945, it is in its current form a fragmented site. The quarry, where prisoners experienced their daily hardship, is not integrated into the memorial site as it is still in use as a quarry. Although visitors could see it from a viewing platform, barely anyone takes this opportunity. Furthermore, the way to the quarry is not an inviting one. It features frequent warning signs about the 'danger to life' due to onsite explosions. Yet once there, the sheer scale of the operation, and most importantly the difficulties the prisoners had to face, is immediately evident.

The adjacent SS buildings of the *Erd-und Steinwerke GmbH* are currently in a state of controlled decay. Whilst Skriebeleit (2016) explained to me in a conversation that there is an ambition to integrate the quarry into the memorial landscape, its current licence agreement prevents it. In addition, significant financial resources would be required to create safe access for visitors. The former SS accommodation, now private residences, will remain outside the memorial landscape. Whilst these houses are outlined on the visitors' map, they are not easily identifiable, and neither are there any markers to indicate that these were SS houses. This is understandable as they are first and foremost private residences; on the other hand, the houses represent the typical Nazi propaganda, a good educational resource. They were built on a hill overlooking a Bavarian valley across to the ancient fortress of Flossenbürg. The houses also feature Bavarian characteristics: wood cladding and hunting trophies. Essentially, they demonstrate the German folklore the Nazis so admired and celebrated. And whilst the camp was on the edge of the village, the SS houses were built alongside the existing village. Therefore, SS staff had a significant presence onsite. In essence, visitors still see only a small section of the former concentration camp landscape.

3.1.3 Flossenbürg as a tourist destination

The hilly landscape of the *Oberpfälzer Wald* encouraged the local authority as early as 1951 to exploit its potential for tourism by attempting to build a ski-lift next to the Valley of Death. Although this project did not come to fruition, Flossenbürg's tourism potential was actively promoted, leading at times to obscure initiatives such as a postcard in 1966 that shows the concentration camp memorial with the message 'Greetings from Flossenbürg'. Flossenbürg's landscape and the ruins of the medieval castle were undoubtedly attractive features, but it was Flossenbürg's proximity to the Iron Curtain, close to the borders of Czechoslovakia and the GDR, that proved to be a magnet for visitors. Large numbers of day tourists were keen to visit the border, behind which lurked the 'communist evil' (Skriebeleit 2009). Nevertheless, for the local authority the strategic development of tourist infrastructure was also an opportunity to fend off the stigma of the concentration camp. A campsite was built next to the local lake *Gaisweiher* which resulted in Flossenbürg becoming the winner of the national competition of 'attractive campsites in the countryside' in 1980, and it finally also received the much-desired label *Staatlich anerkannter Erholungsort* (state-approved health destination).

Ironically, the fall of the Iron Curtain had a negative impact on Flossenbürg. Whilst it had previously enjoyed additional financial support as a *Zonenrandgebiet* (areas close to the inner-German border), with the German unification it was no longer eligible; instead the neighbouring authorities in the newly established state of Saxony now benefited from new investments. Tourism to Flossenbürg has steadily declined since then, and like its counterparts in rural Eastern Germany, Flossenbürg suffers from a population decline. Despite these difficulties, new initiatives like nature tourism as part of the newly formed nature park *Oberpfälzer Wald* are designed to highlight the beautiful landscape and no longer shy away from mentioning the concentration camp memorial.

Derek Dalton (2019) argues that sites such as Flossenbürg provide few opportunities for Holocaust tourism due to its absence of historical features, an argument I do not support. It is indeed the memorial site that now attracts most visitors to Flossenbürg, and the continual popularity of the theologian Dietrich Bonhoeffer leads to pilgrim-like tours. Moreover, as I will show in Chapter 4, it is Flossenbürg's peaceful surroundings that can make it a more interesting location to visit than the busier sites of Dachau or Buchenwald.

3.2 The development of Ravensbrück Concentration Camp memorial

Ravensbrück concentration camp was built near the town of Fürstenberg, 50 miles north of Berlin, taking its name from the nearby village. The village itself had a few settlements and was close to Lake Schwedt. The grounds of the future concentration camp were, therefore, largely uninhabited and mainly used for agricultural purposes, including wood production. In addition, the lake had an important recreational function with three beaches used for swimming in the summer months. Ravensbrück's secluded location, yet near major trunk roads, seems to have been the deciding factor in building a concentration camp for women. The construction work commenced in 1939 using male prisoners from the nearby concentration camp Sachsenhausen, and the first female prisoners arrived in the Spring 1939.

Initially, the largest group of prisoners were Germans and Austrians, one third of them Jehovah's Witnesses (Beßmann and Eschebach 2013). The second largest group were the so-called 'asocials'. At the outbreak of the Second World War, the prisoners' demographic changed significantly. Jewish prisoners were now 15 per cent. However, by far the largest group of prisoners were activists who had been involved in the resistance movement. Some of the most well-known prisoners were Käthe Niederkirchner (communist activist), Corrie ten Boom (a religious activist who became world famous for her books) and Rosa Thälmann (the wife of the Communist, Ernst Thälmann). Prisoners were from across Europe, including France, Italy, Spain, Norway, Denmark and the Soviet Union. Children often arrived with their mothers at Ravensbrück, or were born onsite. According to current knowledge, approximately 880 children between the ages of 2 and 16 years were imprisoned at Ravensbrück.

The women were forced to work in the local textile company, Texled, at the Siemens factory or as prostitutes at Ravensbrück or other camps (Beßmann and Eschebach 2013). They were also subject to medical experiments which consisted of forced sterilisations or Sulfanomid experiments. Sulfanomid as a chemical was thought to avoid infections from wounds soldiers had sustained, yet it had not been tested. Ravensbrück became a medical testing ground for the chemical. Predominantly Polish women had wounds inflicted on them, for example implanting foreign objects such as glass which were then treated with Sulfanomid. Often the women did not survive these procedures or were scarred for life. These medical experiments featured heavily during the British and French war crimes trials between December 1946 and July 1948 in Hamburg.

Between June 1944 and December 1944, a new wave of prisoners, more than 52,000 women, arrived at Ravensbrück (*ibid*). Overcrowding caused dramatic scenes. Food and accommodation were no longer available for these new prisoners and consequently diseases such as dysentery and typhus spread fast. Moreover, the management team decided to erect a temporary tent with the most appalling living conditions. In January 1945, a section of the youth protection camp Uckermark was cleared to make room for women who were destined to die. The women were living in the most inhumane conditions and were provided with minimal food rations that ultimately caused their death.

In the later periods of Ravensbrück's operation, the camp was changed from a forced labour camp to an extermination camp by constructing a gas chamber (Köhler and Plewe 2001). Due to the shortage of building materials, the gas chamber could not be completed prior to the arrival of the Red Army, and consequently a barrack near the crematorium was used from February to April 1945. With the advance of the Red Army, the SS began the destruction of evidence, including the gas chamber, which was demolished. Today, one can only estimate the exact location of the gas chamber as there are no physical remains. This example demonstrates how difficult it is to determine 'historical authenticity' due to the destruction of evidence and the lack of coherent architectural drawings.

In Spring 1945, the International Red Cross negotiated the release of 300 French prisoners who were able to return to France (*ibid*). The Swedish Red Cross received the agreement in February 1945 to repatriate Scandinavian prisoners from German camps. On 8 April, 'White Buses' arrived at Ravensbrück rescuing 100 Norwegian and Danish prisoners. Subsequent rescue missions enabled French, Polish and Belgian prisoners to leave the camp. As the buses were insufficient, these prisoners were allowed to take the train. The missions of the International and Swedish Red Cross saved the lives of 7,396 female and 14 male prisoners.

The Soviet army took possession of the camp on 2 May 1945 and concentrated on the medical care of the remaining prisoners (Beßmann and Eschebach 2013). Officers were also confronted with a high number of corpses at the former Siemens factory and forced local people from Fürstenberg to clear up the camp. In June 1945, the Soviet repatriation camp 122 was established at Ravensbrück, dealing with Soviet prisoners who had been displaced during the war. This camp was closed by the end of 1945 and Ravensbrück became a Soviet military army base, the second largest in the GDR.

3.2.1 The development of Ravensbrück as a memorial 1945–1980

As early as 1948, annual reunions commenced at Ravensbrück, with the memorial consisting of a simple wooden pillar and a 'fire bowl' erected next to the crematorium. After negotiations with the Soviet army, a piece of land was handed over to the GDR authorities in June 1956 for the development of a memorial. In the meantime, the GDR artist Will Lammert had already been asked to propose a structure for the site. He suggested involving the former solitary confinement block and the execution area within the memorial. In addition, a sculpture of a woman was to be erected overlooking Lake Schwedt towards Fürstenberg.

Much like Flossenbürg, an experiential path was designed that would lead the visitor from suffering to enlightenment. The start of the path was a pedestal with a quote from Anna Seghers, a GDR author. The path would then lead the visitor to the 'areas of suffering': the execution path, the crematorium and the cell block. The cell block (solitary confinement) was transformed to house the *Lagermuseum* under the theme *Ravensbrück warnt: Für Frieden in der Welt! Gegen Krieg und Faschismus* (Ravensbrük warns, for peace in the world, against war and fascism). Next to the crematorium, individual nations were commemorated with their names mounted on the former camp boundary wall (Schwarz and Steppan 1999). In front of the wall, a bed of roses was planted at the location of the former mass grave. The roses were donated from various former prisoners' groups including Luxemburg, Hungary, Yugoslavia, France, the Soviet Union and Denmark. From the mass grave, the path would take the visitor to the light, symbolised by Will Lammert's statue *The Burdened Woman*.

The Burdened Woman portrays a woman carrying another woman in her arms on a 25-feet-high plinth looking towards Fürstenberg, with one step forward as if she was attempting to walk across the lake. Referred to as the Pietà of Ravensbrück, it intends to represent the true story of a prisoner carrying another prisoner back to the barracks after she collapsed during an *Appell* (Jansen 1959). With its location at the bottom of the lake, *The Burdened Woman* created the link between the atrocities committed in the camp and the women's yearning for a life in freedom, symbolised by the nearby town of Fürstenberg. In fact, the very slim vertical form speaks directly to the only other high structure in an otherwise very flat landscape: Fürstenberg's church tower.

Thus, *The Burdened Woman* established a permanent relationship between the camp and the town of Fürstenberg which cannot be severed since it is highly visible for residents and tourists alike (Lammert 1965). Unlike in Flossenbürg where a strict boundary was created to separate the local village from the memorial, at Ravensbrück the landscape design created a permanent link to the nearby town.

Will Lammert's statue differs significantly from the heroic, antifascist interpretation of the rest of the site (and at the other GDR's memorial sites Buchenwald and Sachsenhausen) and met with less than overwhelming approval (Peters 2015). Having lost his art teachers in Auschwitz and Majdanek, Lammert was acutely aware of the atrocities committed in the concentration camps and did not subscribe to the antifascist narrative. Lammert's aim was to emphasise the women's dignity despite being surrounded by horrific atrocities. Thus, *The Burdened Woman* features a strong, upright face which tends to be interpreted as looking into the future, yet considering Lammert's personal experiences and the numerous interviews he conducted with survivors, it is more likely to represent the strength and solidarity of Ravenbrück's women.

The *Lagermuseum* exhibition in the cell block, however, followed the GDR's narrative. Günter Meier (1960), an employee of the ministry of culture in the GDR, reflected in the GDR's museum journal on the design of the exhibition. According to him, due to a lack of artefacts, artists had to be deployed to decorate the walls at the exhibition (*ibid*). All interpretation panels used the colour scheme of black, encouraging a cold, hateful atmosphere. Text on the interpretation panels was written using a typewriter setting, creating the illusion that all exhibits are 'official' documents, therefore instructing the visitor to understand the facts. The exhibition would awaken visitors about the courageous fight of the women against the fascist regime. The ground floor of the former solitary confinement block contained reconstructed cells designed to show the torture women experienced, while the upper floor featured national memorial rooms which highlighted the international dimension of the communist struggle at Ravensbrück. Consequently, women were only commemorated collectively according to their nationality and as resistance fighters. Individual victim groups such as Jewish prisoners, Sinti and Roma or the Jehovah's Witnesses merged into a uniform mass of antifascist fighters (Eschebach 1998).

The Burdened Woman was supposed to be surrounded by a group of women at the bottom of the statue who signified the solidarity between the different nationalities and age groups at the camp, yet due to the premature death of Lammert in October 1957 this design was never realised. The change was not accepted by the committee of former Ravensbrück prisoners since it needed to be clear that this camp was a women's and children's camp. A letter was sent to Otto Grotewohl, the first prime minister of the GDR, demanding that the original plans should be implemented. He subsequently suggested that another sculpture should be built: a group of mothers; yet the committee had to wait for another seven years before this sculpture was erected near the memorial entrance.

Hence, the first memorial design was based on an interplay between a hateful past and a hopeful future. The conditions in the camp were shown in their most brutal form with the aim of forging a feeling of disgust against the 'fascist imperialistic' system, and ultimately against capitalism. In addition, the focus on the resistance amongst women was designed to legitimise the communist future. After all, the founding myth of the GDR was the victory over fascism by introducing a better communist state. Yet Anna Seghers' quote at the entrance of the memorial complex also emphasises the treatment of female victims: they were not solely communist victims, but also the future mothers who had been lost in the concentration camp. By contrast, the subcamp Uckermark (for girls between the ages of 16–18) did not meet the criterion of the 'gentle female victim'. Most of the young girls in the Uckermark camp were either 'fallen women' or so-called 'asocials' who did not support the image of the strong, unblemished communist mother, so their suffering fell into oblivion.

3.2.2 Ravensbrück's development as a memorial from 1980 to 1989

A significant change in the history of the memorial occurred in the 1980s. The Soviet army, which had been owner of the *Kommandantur*, handed over the building to the GDR. This meant that suddenly 700 m² were available for a new exhibition. Ravensbrück was subsequently closed and the building work for the new exhibition began. The new central exhibition focused on the development of Ravensbrück concentration camp in addition to the organised antifascist fights of the brave women (Unknown 1982). The 1980s museum concept emphasised that 'the exhibits should cause hatred and condemnation amongst all visitors against the atrocities of the fascist German imperialism and should at the same time encourage the fight against the aggressive inhumane politics of imperialism in the present' (*ibid*). The exhibition was then divided into four different sub-themes encompassing the historical development of the camp, a model of the camp, a history of the resistance and finally the legacy of the antifascist fight and its implementation in the GDR. The central exhibition's aim of inducing hatred amongst visitors led to a continuation of the shock pedagogy to an extent that objects and images were exhibited that did not have a connection to Ravensbrück, but East German visitors were kept in the dark about these facts (Eschebach 2015).

Eschebach (*ibid*) notes that in the first few rooms of the 1980s exhibition seven oversized images of women and children were featured behind barbed wire. None of the images were, in fact, from Ravensbrück: two were taken at Auschwitz and one was the well-known image of the young boy in the Warsaw Ghetto. The visitors were not informed of the origin of the image, their context or indeed the year they were taken. Furthermore, much like in the 1960s exhibition, artists hand-painted typical scenes from life in the camp onto walls (Porsdorf 2019, personal communication). In fact, the artists stayed onsite during the design process which was a challenging experience for those involved as Friedrich Porsdorf explained that 'the stay at Ravensbrück haunted me'.

His paintings were also placed behind barbed wire fences which he rejected as it distracted from the message he was trying to convey. However, since the aim of the overarching museum concept was to emotionally engage visitors, the images behind barbed wire fences remained in the museum (Eschebach 2015). Particularly distressing in the exhibition, according to Litschke (1985), the manager at Ravensbrück in the 1980s, was the portrayal of medical experiments on women and children, alongside the systematic killing of Jews and Sinti and Roma. Thus, the 1980s exhibition extended the concept of shock pedagogy by playing with visitors' emotions. However, Litschke's mentioning of Jews and Sinti and Roma highlights a slight shift in the GDR towards the end of the 1980s. Since the GDR sought greater acceptance with the West, it began to acknowledge specific victim groups.

Alongside the development of the new museum, the former cell block was also re-designed. Egon Litschke decided to reconstruct the cells of the ground floor including cell doors (Eschebach 2008). He emphasised that the bleak and cold atmosphere of the cell block and the echo of the steps should evoke an emotional reaction. On the upper floor of the cell block the individual national memorial rooms were also re-designed, either by the nations themselves or by the GDR's Ravensbrück management team with the limited information they had. This design of the cell block exhibition is still visible today.

3.2.3 Ravensbrück's development as a memorial from 1990 to the present day

The politicisation of Ravensbrück's history to fit the GDR's political agenda was deemed inappropriate after the fall of the Berlin Wall. In line with the national Enquete Commission, in 1991 the federal state of Brandenburg launched a separate expert commission for the re-development of the memorial sites. This resulted in the formation of a separate trust which would be responsible for the memorial sites of Ravensbrück, Sachsenhausen and Potsdam-Leistikowstraße. Brandenburg's expert commission recommended that the GDR exhibition was to be closed and replaced by a new exhibition called *History and Topography of the Women's Concentration Camp* (Ministerium für Wissenschaft und Kultur Brandenburg 2009). In addition, it recommended that the former Youth Protection Camp Uckermark, the Siemens factory, the former SS guard houses and the wider grounds of the camp should be incorporated into the memorial. However, this was a strenuous undertaking. The Soviet army left behind a wasteland: the grounds were contaminated by chemicals such as kerosene, and the former camp structures had been significantly altered.

A landscape architectural contest was launched in 1998 with the aim of transforming the site, much like Flossenbürg, bringing it closer to its 1945 appearance without reconstructing buildings. The architectural bureau Oswalt/Tischer won the overall contest. Their proposal included the covering of the former area of the barracks with a surface made of clinker, creating a relief of the outlines of the former barracks. If the Soviet army buildings disturbed the memory of the concentration camp they would be demolished. The former camp pathways would be marked by trees, whilst any other vegetation which had grown onsite since 1945 would be removed. As there was no physical evidence left of the former Uckermark camp, the ground would be planted with wildflowers to demonstrate the fragility of life (Oswalt and Tischer 1998). The first part of the landscape design, the surface made of clinker, has been completed, permitting visitors access to the area of the barracks for the first time. This created a vast and bleak open space where no vegetation is allowed to grow. While it was designed to resemble the suffering of the victims, Oswalt and Oswalt (2000) acknowledge that without further explanation, visitors would be unable to interpret the significance of this design.

Returning the site to its 1945 appearance has removed almost all traces of the former Soviet army base, a 49-year history (between 1945 and 1994). One could, therefore, argue that this is a new form of selective memory, since the voice of the Soviet soldier who lived onsite is erased. In fact, the landscape design team Oswalt and Tischer envisaged retaining the traces of the Soviet army; yet this was rejected by the memorial's management team, survivors and the local conservation authority (Tischer 2020, personal communication). Neither the Uckermark camp nor the former Siemens factory was integrated within the current memorial landscape. With regard to the Uckermark camp, this was due partially to the severe contamination of the land and partially to difficult ownership arrangements: that is, the former Uckermark area was not owned by the memorial trust (Oswalt and Oswalt 2000). Today, the Uckermark camp is theoretically accessible for visitors, yet is not on any official visitor map. It is approximately 3 km from the main camp and situated on the new Berlin-Copenhagen cycle path. Unless a visitor ventures off the cycle path s/he will not access this area. The area itself is currently maintained by political activists who visit the site approximately once a year for general maintenance. The Siemens factory is only accessible by an adventurous climb over fences which visitors, of course, are not allowed to undertake.

In 2006, yet another overarching memorial concept for Brandenburg's memorial sites, *Geschichte vor Ort* (History Onsite), was developed. The main exhibition in the *Kommandantur* would be the central focal point surrounded by the smaller, more detailed satellite exhibitions at the *Führerhaus*, the SS Female Guard House, the Cell Block and the Textile Company (Ministerium für Wissenschaft und Kultur Brandenburg 2009). While most stakeholders approved of the overall concept, the local Ravensbrück committee criticised the exclusion of the Uckermark camp and the former Siemens factory again. During my research in 2016, the SS female guard house was closed, so I could not capture visitors' responses to this challenging exhibition, but I will nevertheless highlight here the controversies surrounding the exhibiting of women as perpetrators.

3.2.4 Exhibitions at Ravensbrück

The visitor encounters the memorial complex usually from the main road in Fürstenberg. S/he then first sees a group of women symbolising the future mothers lost at Ravensbrück. This statue was created by Fritz Cremer after Otto Grotewohl responded to a complaint by former prisoners about the neglectful representation of the mothers in the camp. Continuing on the main path to the memorial, visitors are confronted with a Soviet army tank which represents the liberation of Ravensbrück by the Red Army alongside remnants of former SS Guard houses and Soviet army accommodation. On arrival at the car park, visitors are greeted with a new visitor centre opened in 2007, opposite Lake Schwedt. There is no requirement to visit the centre unless one wants to obtain a map or an audio guide. The overarching memorial concept envisages visitors going first to the newly-opened exhibition in the *Kommandantur* before exploring the additional in-depth exhibitions.

The new exhibition 'History and Memory of the Women's Concentration Camp' in the main *Kommandantur* was opened in 2013. For the first time, a more comprehensive perspective of the under-researched female experience of a concentration camp is exhibited, based on the conditions of the camp, the various victim groups, medical experiments and forced labour. Furthermore, the marginalised victim group of the 'asocials' in the Uckermark camp is highlighted. As such, the curatorial team attempted to show multiple perspectives of victimhood in the exhibition. For instance, previously, the portrayal of victim groups was often based on solidarity and mutual support, yet a publication by the Austrian Victim Association in 1945 highlights that this is not a universal truth (Bruha et al. 1945). The communist victims condemned the so-called 'asocial' women, and the women often retained this stigma after liberation.

Ravensbrück placed a greater emphasis on cultural objects made by the women in the camp instead of displaying typical relics (e.g. shoes or striped uniforms). Those objects are often representative of individual acts of resistance, expressions of hope or symbols of one own's identity (Rydén 2018). Johanna Rydén (*ibid*) who studied the Ravensbrück collection housed at Lund University, Sweden, which includes a collection of items former prisoners donated after being rescued by the Swedish and Danish Red Cross, explains that most objects were delicate gifts of beauty and affection that assisted in the fight for survival but also strengthened solidarity between the women. As such, Ravensbrück's material culture differs significantly from those objects displayed at other camps such as Auschwitz or Sachsenhausen. Chloe Paver (2018) highlights that the focus on *Lagerkultur* (daily life in the camp) signifies a caesura in exhibition design culture as planners moved away from the sole display of 'trauma icons'.

The most challenging part of the new exhibition is the section about sex slave labour. SS commander Heinrich Himmler introduced brothel systems at ten major concentration camps as an incentive for male prisoners. Ravensbrück hosted one of those camp brothels, yet also operated as a 'supply chain'. For women, it was a double victimisation. Not only were they subjected to arbitrary imprisonment, but also forced to work as prostitutes. This aspect of the camp's history is exhibited using a variety of eyewitness accounts, including those of the male prisoners who used the brothel. When Ravensbrück first acquired the exhibition on sex slave labour it collaborated with art students who identified words which were used in the discourse about sexual violence at concentration camps. Subsequently, single German words such as 'Schlampe' (slut), 'Asozial' (asocial) and 'Freiwillig' (voluntarily) were projected onto a wall (Paver 2018). Furthermore, comments made by inmates about their perception of the women highlighted the attitude some male prisoners held. Images were not shown in order to avoid 'sexploitation'; the connection between Nazi institutions and sexual violence which is repeatedly shown in films about women's prisons (Eschebach and Jedermann 2007). Today's permanent exhibition at Ravensbrück focuses almost exclusively on the victim's perspective, thus providing a less challenging exhibition for visitors (Paver 2018). In fact, Matthias Heyl (2016), head of education at Ravensbrück, explained that visitors still often leave with the impression that the brothels were introduced for SS staff.

Alongside the remnants of mass death, the issue of perpetration is confronted in two different exhibitions. One of the former SS guard houses was transformed into an exhibition about the female staff at the camp. Ravensbrück, therefore, addressed a taboo topic: women as perpetrators. Displaying female SS perpetrators was, and still is, not without controversy. Before the opening of the first exhibition, significant debates were centred, for instance, around the display of a private photo album that showed the SS officer as an ordinary person. It challenged the stereotypical narrative of the SS officer as brutal male. In 2020, a new exhibition was opened under the theme *Akzente setzen – Farbe bekennen* (Emphasising features – coming clean), whose aim it is to represent the new historical knowledge about female SS guards, but also to show their often (sexualised) portrayal in popular culture after 1945. In so doing, Ravensbrück wants to exhibit the 'grey areas' of the female perpetrators, often ordinary women who, however, turned violent very quickly.

In order to avoid the perception that only female SS staff worked at Ravensbrück, the memorial team also designed an exhibition in the former *Führerhaus* that addresses the male SS staff. All SS guard houses, including the *Führerhaus*, had been converted into officers' accommodation during Ravensbrück's time as a Soviet army base. Hence, the *Führerhaus* had to be altered to show the original layout. However, a complete reconstruction was not carried out. On entering the *Führerhaus*, one can initially hear the voices of liberated prisoners who were allowed into the house for the first time. The visitor can then explore the former kitchen, the bathroom and the bedrooms. The main aim of this exhibition is to display the contrast between the luxurious lifestyle in the SS houses and the inhumane conditions in the camp only a few metres away. In addition to the biographies of the SS male management team, the wives of the officers, including the

Kommandant's wife, are also displayed. These wives were often supported by maids from the camp, so the intention in the exhibition is to emphasise the collaboration of the women, hence creating again the link to the female perpetrator.

Another part of the memorial is the former GDR memorial complex with its solitary confinement block, the crematorium and the memorial wall. After controversial discussions, the international committee of the memorial trust decided in 2003 that no single room of the GDR's at times controversial exhibition should be altered or removed (Eschebach 2008). Consequently, the visitor will encounter the former GDR exhibition with its socialist propaganda. Particularly difficult are the memorial rooms of the Soviet Union and Bulgaria. The Soviet memorial room consists of a range of images that do not have a connection with Ravensbrück. In fact, they document the atrocities committed by the SS in the former Soviet Union. The 14,000 prisoners who had been deported to Ravensbrück due to minor offenses were neglected. With the collapse of the Soviet Union, the newly formed states such as Belarus, the Ukraine or Latvia reject this memorial room (*ibid*). The Bulgarian memorial room was designed in 1987 with the message 'With dedication, the Bulgarian women join in the armed antifascist struggle of the Bulgarian people' (Eschebach 2008, 113). Thus, most images display women of the first labour movement in 1912 and therefore represent the antifascist resistance. Yet the most disconcerting part of the room is the portrayal of women in daily life in Bulgaria in the 1980s, symbolising the communist achievements. The room is so controversial that the *Frankfurter Allgemeine Zeitung* reported on 26 April 2004 that Bulgarian Socialism is alive and well in Ravensbrück's solitary confinement block (*ibid*).

A visitor to today's Ravensbrück has to effectively negotiate two memorial complexes: the GDR's design with its Christian experiential path from the dark to the light and contemporary Germany's design with a focus on the past. Like Flossenbürg, the current memorial represents a fragmented version of the former concentration camp. The areas of severe suffering, for example the location of the tent, the Siemens factory and the Uckermark camp, are not part of the current presentation. Indeed, Ravensbrück's memorial landscape is largely characterised by absence. Hence, the positioning of the visitor within the wider memorial is crucially important. It is his/her imagination that will create the meaning of the visit (Violi 2012). Thus, Ravensbrück is also symbolic for the difficulties of managing 'traumatic' landscapes dominated by absence. Natalie Bormann (2018), for instance, highlights that Ravensbrück has a ghostly appearance with its vast empty spaces. Therein, however, also lies the challenge. How does one interpret and manage landscapes which were witnesses to past human suffering, an issue that has so far received little recognition (Rapson 2015). Yet with the expansion of the memorial site to include the south side (the location of the tent), now a grassland, such questions will become ever more pertinent (Starke 2017).

3.2.5 Ravensbrück within the wider tourism landscape

Unlike Flossenbürg, Ravensbrück and its neighbouring town of Fürstenberg have always been part of a thriving tourist industry. Located in the picturesque area of the North German lake district *Mecklenburger Seenplatte*, Fürstenberg had its heyday in the 19th century when wealthy Berliners used the town for spa and recreational activities. Lake Schwedt, in particular, was an important leisure facility, which the construction of the concentration camp changed for good. With the Soviet army taking possession of the former concentration camp, the lake was now almost off limits for tourists and residents alike. Moreover, most of the lakeside properties were transformed into housing for Soviet Officers and were no longer available as tourist accommodation.

Nevertheless, unlike at Flossenbürg, the GDR actively promoted Ravensbrück as a destination. The memorial site was mentioned in all GDR's tourist guides, including an often comprehensive description of its history and the onsite museum. Ulrike Helwerth (1990) writes that at least six school classes used to arrive at Ravensbrück, and numerous annual commemorative events used to take place, thus making Ravensbrück a much-visited memorial site in the GDR. Yet with the fall of the Berlin Wall, Ravensbrück changed into an eerie memorial site. Since visits to concentration camp memorials are no longer a compulsory part of the German curriculum (except for Bavaria), visitor numbers dropped significantly.

However, Ravensbrück benefits from a thriving largely domestic tourist industry. The *Mecklenburger Seenplatte* with its canals and lakes is an attractive destination for water sports enthusiasts. In fact, there have been frequent demands to create a boat landing stage at the bottom of the statue the *Burdened Woman*, which was rejected as a trivialisation of the site (Eschebach 2011). On the other hand, a section of the former female SS guard houses was transformed into a youth hostel, allowing visitors to stay overnight. Moreover, Germany's first 'e-bike highway', a long-distance cycle path with e-bike charging stations that connects Berlin with Copenhagen, runs adjacent to the memorial sites, thus attracting day trippers and cycling tourists. The tourist information centre and sightseeing tours on the Lake Schwedt mention the concentration camp memorial, so Ravensbrück is firmly embedded within the wider tourism infrastructure.

3.3 The House of the Wannsee Conference

The Haus am Großen Wannsee 56–58 has a colourful history. It was originally built in 1914/15 for the wealthy industrialist Ernst Malier, who claimed that he found a cure for obesity and insomnia, which made him enough money to build a house in the popular Wannsee area (Digan 2014). Unfortunately, it was soon evident that his cure did not achieve the desired results, and Malier was accused of fraud. He then sold the house to the *Norddeutsche Grundstücks Aktiengesellschaft*, which went bankrupt in 1937 and whose only remaining stakeholder, Friedrich Minoux, inherited the house. However, he was accused of having misappropriated eight million Reichsmark from the Berlin gas company in 1940. Minoux was subsequently sentenced to five years in prison. Due to the villa's beautiful surroundings, the Stiftung Nordhav (set up by Reinhard Heydrich to acquire holiday homes for SS members) bought the house in 1941 as a conference and guest house. A few years later, on 20 January 1942, the Wannsee Villa became the ideal location for a meeting of 15 high-ranking representatives of the SS, the NSDAP and various ministries to discuss their cooperation in the planned deportation and murder of the European Jews, now known as the Wannsee Conference.

In March 1947, whilst collecting information for the Nuremberg trials, the US protector came across a document he would later claim was the most shameful document in history: the *Wannsee Protokoll* (the German word Protokoll corresponds to the English word 'minutes') (Roseman 2002). The minutes are largely presented by Heydrich and detail the Jewish situation up to this point whilst also considering the final extermination. The participants of the conference were representatives from ministries which had responsibilities for the 'Jewish question', followed by representatives from ministries from the Eastern Occupied Territories. SS functionaries with an interest in race questions also participated.

The conference commenced with Heydrich explaining the reason for the meeting: to plan the final solution for the 'Jewish question' (Digan 2014). Suggestions were made on how to exterminate the Jewish population across Europe. All in all, it was a 'successful' day as all representatives responded positively to the suggestions. Even Stuckart (state secretary in the German Interior Ministry), who had previously been hesitant, seems to have agreed, according to the Protocol. The Wannsee document itself remains puzzling and is often at the centre of heated debates around the Holocaust. At the time of the conference, the murder of Jews had already begun, and the Belzec extermination camp was under construction (Roseman 2002). So what, therefore, was the purpose of this meeting, especially as Hitler was not present? Historians have struggled with an explanation to this date, with the German historian Eberhard Jäckel commenting in 1992: 'The most remarkable thing about the Wannsee Conference is that we do not know why it took place.' Today, however, it is widely accepted that the Wannsee Conference was called to facilitate the careful planning of a mass genocide.

3.3.1 A memorial? The debate surrounding the 'Haus am Wannsee'

After the Second World War, the villa could not be returned to private ownership and, therefore, the council of Berlin took possession of the house (Kühling 2008). It is not known how the house was initially used, but in 1947 the Social Democrats became tenants of the villa and established a training centre, the August Bebel Institute. However, the institute ran into difficulties in the 1950s and vacated the house in 1952. The villa was then leased to the Berlin district of Neukölln who used it as *Schullandheim* (country home for schoolchildren).

While the centre was in operation, it could accommodate two or three school classes of up to 40 children. Interestingly, although the brochure of the *Schullandheim* detailed the history of the villa, it never mentioned its association with the Wannsee Conference.

The historical importance of the site was largely forgotten until 1966, when Joseph Wulf, an Auschwitz survivor, started a campaign to establish a documentation centre at the original Wannsee villa. Wulf, who called himself the 'the tattooed man of Auschwitz', swore on his liberation that he would dedicate his life to the history of the Third Reich. At the heart of his dedication was the formation of a trust and the establishment of a memorial at the iconic Wannsee house. He also convinced Nahum Goldmann, the president of the World Jewish Congress at that time, to support his project. Initially, conversations with Egon Bahr, the press officer of the Berlin Senate, were positive. However, by the time Goldmann visited Willy Brandt (the mayor of Berlin) in October 1966, the press in West Berlin had begun a negative debate in the public sphere. The media seemed to be outraged by the fact that the children of Neukölln should suffer in favour of the erection of a memorial. This school of thought continued with Klaus Schütz (the new mayor of Berlin), who apparently said in 1967 that he hoped that the cheerful children's laughter in the *Schullandheim* would finally drown out the death cries of gassed Jewish children. Another concern mentioned in the press was the potential rise of antisemitism if a memorial site were to be established that focused on the perpetrator.

Support for the new memorial came largely from the Jewish community. For this community, the fate of the Jewish people during the Second World War was intrinsically tied to the Wannsee house. Wulf argued that the villa had a unique character and should find a use fitting to its character (Digan 2014). *Der Tagesspiegel* reported in 1966 that while legally the house belonged to the Senate, morally it belonged to all countries that were occupied by Germany (Kühling 2008). The debate about the House of the Wannsee Conference was typical for Berlin in the 1950s and 1960s, and West Germany as a whole. Memorial sites in Berlin focused mainly on the resistance, and the Berlin Senate justified its negative view in terms of the Wannsee villa by highlighting the memorials at Stauffenbergstrasse and Plötzensee. The active engagement with the Holocaust (although not called the Holocaust in these early years) was a Jewish concern and not an issue Germans needed to deal with. Consequently, Schütz sent a letter to Wulf on 20 December 1967, explaining that the Senate had decided that the villa would not be available for a documentation centre. The battle for a memorial was lost.

The Senate offered two alternative buildings near the Freie Universität Berlin. However, this was unacceptable to Wulf, his Trust and Nahum Goldmann (*ibid*). Nevertheless, the buildings were viewed, and it turned out that they were not available after all. In May 1970, the Senate offered a building associated with a research centre at the Freie Universität Berlin, but again this was never implemented. In September 1970, Wulf gave up and resigned as a president from the Trust he had formed. He had been unable to achieve his lifelong goal which led to him committing suicide in 1974. In his final letter he wrote:

I have published eighteen books here on the Third Reich, but this had no impact. You can publish things for the Germans until you are blue in the face, they might be the most democratic government in Bonn, but the mass murderers wander about freely, have their little houses and cultivate flowers.

(Später 2013)

After Wulf's death the memory of the Wannsee Conference disappeared from public consciousness. Although a memorial plaque was erected in 1972, it was vandalised and not replaced until 1982 (Kühling 2008). The commemorations on the 40th anniversary, led by the head of the Jewish community and the mayor Richard von Weizsäcker, brought the villa back into public memory. Another attempt was made to develop a memorial site, yet again it failed.

A change of thought happened in 1985. The Berlin Senate finally decided to develop the Wannsee Villa as a memorial site. However, they did not want a 'dead' museum. The intention was to create a space for commemoration, research and international dialogue. A new trust was formed that was responsible for the design of the first exhibition. The site was finally opened in 1992. At the opening ceremony, Eberhard Diepgen (Mayor of Berlin) commented that the Wannsee memorial would finally confront the Germans with their dark history. Together with the memorials at the former Prinz-Albrecht-Palais (Topography of Terror) and the memorial for the German Resistance at Stauffenbergstraße, the Wannsee memorial would form a triangle of remembrance. Wulf could not have imagined this outcome when he campaigned for the Wannsee memorial in the 1960s.

3.3.2 The development of the exhibitions 1992–2020

The first exhibition was developed between 1988 and 1991 and officially opened on 20 January 1992, the 50th anniversary of the Wannsee Conference (Geißler 2015). This exhibition was shown until 2005 with few alterations. The responsible curators, Gerhard Schoenberner and Wolfgang Scheffler, decided that the Wannsee Conference would be the core of the exhibition, yet it would not dominate it, leading to criticism in the early years. The exhibition was a 'silent exhibition', that is, no interpretative intervention (audio-guides, interactive exhibits) was used, and the focus was on large photographs and some carefully selected documents. Scheffler argued that this exhibition concept and the deliberate use of shocking photographs would finally force the Germans to face up to the Nazi atrocities (*ibid*). In 2002, the Trust decided to review the exhibition and concluded in 2004 that the exhibition needed to be re-designed as it no longer met the expectations of modern museum education techniques. In addition, the opening of the archives in Eastern Europe allowed new insights into the planning and execution of the genocide, which needed to be incorporated into the new exhibition. Interestingly, the Wannsee memorial also had to re-position itself within the memorial landscape of Berlin. With the new Holocaust memorial in the centre of Berlin, Wannsee was no longer the only site that commemorated the mass murder of the Jewish community.

The new exhibition under the title 'The Wannsee Conference and the Genocide of European Jews' opened in 2008 and is, as Norbert Kampe (head of Wannsee memorial from 2004 to 2015) explained, a perpetrator site (Pearce 2011). Rather than the shocking photographs of victims of the previous exhibition, biographies of four families in relation to the Holocaust were portrayed. Their stories reappeared throughout the exhibition when various stages of the Holocaust were explained: deportation, ghettos, forced labour and concentration camps. There was a noticeable absence of perpetrator voices, as they often never admitted their crimes. Hence, the perpetrators at Wannsee were portrayed using the story of Nazi bureaucracy.

Since the visitor research that underpinned this project was based on the second exhibition (2008–2020), I will explain its design and layout in more detail. The exhibition was designed as a circular path through 15 themed rooms (Geißler 2015). The visitor entered room one through the foyer of the villa which contained an introduction to the exhibition, followed by a detailed description of the development of the race ideology during the Weimar Republic and in Germany between 1933 and 1939 (themed rooms two to four). Rooms five and six explained the Second World War, featuring the examples of Poland, Serbia and the Soviet Union. The following rooms (seven to ten) explained the Wannsee Conference, its significance and the people and institutions involved in the Holocaust. The subsequent conservatory described very briefly the history of the house, but its main focus was the Eichmann trial, which visitors could listen to via audio stations. The former kitchen area (rooms 11 to 14), a very dark area with no windows, showed the stages of the Holocaust. The exhibition ended with a presentation under the title 'The Past in the Present' in which individuals talked about their post-war experiences (e.g. the great niece of Himmler talks about her life experiences).

Throughout the exhibition, the visitor encountered the biographies of different Jewish families. The families were introduced in room one: Albert Silberstein from Berlin, Alexandre Halaunbrenner from Paris, Esther Reiss and Eugenia Tabaczynska from Poland. At the beginning the Silbermans are shown in their home in Berlin-Lichterfelde as a happy and successful family. The visitor encountered the Silbermans again in room three (Jewish population during the Weimar Republic), room four (Jewish persecution), rooms 12 and 13 (concentration camps) and room 15 (the past in the present). The same exhibition principles applied to the other families shown in room one. The personal fate of these families was exhibited using photographs, testimonies and, in the case of Esther Reiss, an audio station with her experiences of the ghetto in Lodz and the concentration camp Bergen-Belsen. The aim of the concept was to show the perpetrators' actions and long-term consequences for these Jewish families. This led to a fragmentation of their biographies and, as Cornelia Geißler (2015) argued, it was unclear how the visitor should have established a connection between the different fragments. Indeed, the personal histories were buried within the very factual exhibition and, unless the visitor sought them out, s/he would not have learned about the fate of these families.

In many respects, the introduction of personal stories felt like an afterthought. This type of exhibition design was rooted in Germany's new approach to museum pedagogy: with the generational change, a personalisation of history was thought to be a better way of displaying the Nazi past rather than showing *Leichenbergfotos* (photos of piles of corpses) (Brink 1998). Furthermore, as Elke Gryglewski (2016) pointed out in interview, some photographs induced secondary victimisation (as these images were often made by the perpetrator) or emotionally overwhelmed visitors, a pedagogical strategy no longer deemed appropriate at German memorial sites under the *Beutelsbacher Konsens*.

In room nine, the location of the Wannsee Conference, the victim perspective was completely absent. This room focused entirely on the actual conference, its participants and the Protocol. The Wannsee memorial management team deliberately did not reconstruct the room as they were keen not to 'stage' history. An image of the furnished room provided visitors with a glimpse of the atmosphere of the house at the time. The protocol itself was also a copy because the conditions of the house did not allow the display of the original. The text in this room made it clear that the decision to exterminate the Jews across Europe had already been made by the time the conference took place. The exhibition in the final room 15, 'The Past in the Present', was a space for visitors to reflect on the long-term consequences of the Holocaust. Children of Holocaust survivors talked about their experiences of growing up in the shadow of the Holocaust, whilst children of perpetrators talked about their guilt. Geißler (2015) criticised this concept as being very one-sided, in particular in relation to the perpetrator families, since it did not mention that the topic of the Holocaust was often never discussed in these families. Whilst these personal quotes were certainly often emotional and emphasised the difficulty of living with either survivors or perpetrators, it was, as Geißler (*ibid*) noted, a missed chance to engage with the wider topic of memories within family contexts.

3.3.3 The new 2020 exhibition

Throughout my visitor research detailed in the subsequent chapter, the management team was already aware of the exhibition's shortcomings and so had commenced working on a redesign which was opened in January 2020. The design bureau, Franke in Berlin, was tasked to develop a new exhibition, with a more inclusive design. Initially, the designers suggested recreating the atmosphere of the 'conference with breakfast', using video projections (Franke 2021). This was, however, rejected by the memorial's management team by arguing that the conference could have taken place in any other office building: there was no causal connection between the building and the conference. Instead, the design team worked with colour schemes to change the atmosphere of the rooms. The debate surrounding the 'reconstruction' of an atmosphere using virtual technologies highlights Germany's different approach to exhibition design. The aim of the exhibitions is a factual account of historical events, and therefore the inauthentic recreation of former historical events is considered to be inappropriate. In so doing, Germany does not follow the interactive designs one can so often witness in Holocaust exhibitions outside Germany.

Unfortunately, historical research into the interior of the villa revealed very little information about its appearance in 1942. However, traces found during the renovation exposed a much darker colour scheme than the one used in 2008. Consequently, the former kitchen (room one) is painted in blue, which also forms a contrast to the historic tiles. Indeed, the visitor is now greeted with a visual projection of the invitation to the conference, accompanied by the sound of a typewriter. Thus, the visitor is immediately drawn into the cool and factual atmosphere of the conference and is no longer, as it was in the previous exhibition, firstly informed about the rise of the Nazi's race ideology.

The rather cool colour scheme continues throughout the exhibition, which forms a contrast to the beautiful surroundings of the villa. It was hoped that the exhibits would be free-standing instead of being mounted onto the walls, which would have allowed visitors to approach them from different angles. The large visitor volume prevented this design format, but the sloping lines on the exhibits themselves were kept to signify the landslide magnitude of the moral and human transformations during the Nazi regime. A key aim of the exhibition was to be as inclusive as possible, thus the font 'MS me' was chosen for its readability and friendliness, while for describing perpetrators and Nazi institutions a harsher font was used.

When comparing the new exhibition with the previous one, the reduction of images is immediately evident, in particular those that have shown violent scenes. Due to visitors at times recognising their relatives in the images, and visitors being emotionally overwhelmed, such images are no longer prominently on display; instead, they are embedded within the touchscreen applications or as additional information. Although the Wannsee Conference now takes greater centre stage, the focus on the actual protocol was removed by no longer using a large glass table to exhibit the document. The history of the conference is now also embedded within the wider societal context. Exhibits explain the *Täterschaft* (perpetrators) beyond the main participants of the Wannsee Conference, the systematic exclusion of Jews from public life and its support from wider German society. The final parts of the exhibition take a reflective stance on memory politics in Germany in dealing with the Nazi regime and the Wannsee Conference in popular culture. In fact, one exhibit focuses on the treatment of perpetrators in the Soviet Occupation Zone. For the first time, the previously neglected Eastern German history enters the discourse. By reflecting on German memory politics, the House of the Wannsee Conference follows Flossenbürg's 'What remains' exhibition. Indeed, the memorial now also reflects on itself by providing exhibits throughout the exhibition about the villa's use since 1945. Hence, the memorial site emulates the increasing trend of museums critically reviewing themselves and their relationship to the exhibitions.

Before visitors leave, they are now invited to contribute to the exhibition by sharing their thoughts in a virtual guest book. Entries are instantly recorded and grouped according to headlines such as 'shame' or 'guilt'. By doing so, the visitors' comments become a part of the exhibition. A strong desire of the new exhibition was to increase visitor participation, which has already led to the first scandal. One exhibit was designed to encourage visitors to think about contemporary discrimination. For instance, an image of a swimming pool in Frankfurt/Main was shown which restricted single male refugees from entering the pool for fear they would harass women. Another image showed a large queue in front of a cheap clothing store, hinting at modern-day slavery and exploitation in the textile industry. This approach was, however, criticised for being insensitive to the Jewish suffering. How could the memorial site compare a large-scale genocide with refugees being barred from a swimming pool? Consequently, this exhibit was removed from the exhibition shortly after the opening.

The development of the new exhibition shows the dilemma German memorial managers face. The buildings themselves, in this case the Wannsee villa, do not necessarily lend themselves to modern museum exhibition design techniques. In addition, an international audience demands a variety of different interpretation techniques which cannot be easily implemented within a confined space. Wannsee also has to define itself within the increasing memorial landscape in Berlin and indeed Germany, while also wanting to be true to its original mission: an education and documentation centre. With a stronger focus on participation, the memorial site attempted to involve visitors who do not participate in formal education programmes, yet it was heavily criticised for doing so, which shows Germany's other dilemma; Germany's memory culture with its focus on the singularity of the Holocaust clashes with the desire to engage modern audiences who view the Nazi past as a distant past.

3.3.4 *The House of the Wannsee Conference within the tourism landscape*

Derek Dalton (2019) notes that the House of the Wannsee Conference is the neglected site within the academic literature on tourism to memorial sites, despite its large visitor volume. Within this research project, the House of the Wannsee Conference represented a site that does not feature any physical traces of mass atrocities. The lakeside setting, the beautiful architecture of the villa and the carefully managed gardens of the Wannsee memorial site stand in stark contrast to the concentration camp memorials. Here, there are no traces of mass murder, and yet by its association with the Wannsee Conference, it has a direct link to the genocide of the European Jews.

There are no artefacts at the memorial site, including no furniture from that time. The memorial, therefore, opted for material authenticity with a focus on the house itself. Experiential forms of visitor engagement, for example, the recreation of victims or perpetrators' experiences, aside from audio stations, was not used in 2016 and also does not feature in the new exhibition. Derek Dalton (*ibid*), for instance, wondered during his first encounter with the Wannsee Villa whether an animation of a film scene from the German film *Wannseekonferenz* would fill the void in the house, and give visitors an impression of the SS presence. Yet such approaches are firmly rejected within Germany's concept of memorialisation.

And yet, as Caroline Pearce (2011) suggests, visitors arrive at memorial sites with certain preconceptions. In terms of the perpetrator, there is the mental image of sadistic killers. In fact, Wulff Brebeck (1995) explains that for most visitors the everyday criminality of the Nazi administration is less significant than the horrific crimes committed at the concentration camps. Furthermore, since the opening of the memorial site, an emotionalisation of the site has taken place (Kühling and Jasch 2017). These emotions are heightened when visitors arrive with images of the film *Conspiracy* in their head: for example, Dalton mentions that as soon as he entered the house, he remembered the film. Since visitors do not encounter a site that is furnished in its 1942 appearance, it might disappoint the 'Dark Tourist' who wants to feel as 'it really was' (Jasch 2017, 149). Visitors, for instance, frequently complain about the table's removal in the 2008 exhibition and do not accept that it was an artistic installation.

Thus, Jasch (*ibid*) suggests that educational programmes need to 'break the spell of the dark site of house' by connecting the protocol to its implications. He then sets out Wannsee's extensive education programmes with schools, but also public institutions such as the *Bundeswehr*, the police or the health sector. Yet this methodology has neglected the individual visitor. Already in 1992 staff lamented that the volume of Holocaust tourism would make it impossible to fulfil its educational mission. Indeed, according to Kühling and Jasch (2017), in times of Holocaust tourism where the visitor is only interested in the hype surrounding the perpetrator, a pedagogical offer is increasingly difficult.

Whilst the large visitor volume is certainly an issue at the House of the Wannsee Conference, often leading to overcrowding, I am more concerned about the negative view of the 'hyped-up tourist'. During my research, the Wannsee Conference could not provide an educational offering for individual visitors; in fact, it discouraged families with children from visiting the site. And yet I accompanied several visitors who wanted to have a meaningful visit and were left 'hanging'. Moreover, I witnessed how individual visitors were, in essence, forced out of the room without any explanation when a large educational group entered the space. By contrast, for Jewish tour groups, the House of the Wannsee Conference is often the first stop on their journey to Poland. Their visit was often reduced to the conference room since this was the location where the horror was facilitated, and Wannsee was only a stop on a busy tour operator itinerary. Wannsee has a very diverse audience that ranges from the casual visitor who might chase an authentic representation, to the visitor who seeks a meaningful encounter with a challenging past and various shades in between. Yet these audiences were not catered for unless they had booked a specific educational programme.

3.4 Bautzen II Stasi prison memorial

The town of Bautzen is the location of two prisons: Bautzen I and Bautzen II. Although my research focuses on the Stasi prison Bautzen II, I will briefly explain the history of both prisons as they are intertwined. Bautzen I, known as the Yellow Misery due to its yellow facade, was built on the outskirts of Bautzen in 1904 for 1,100 inmates. Between 1945 and 1950 it functioned as a Soviet Special Camp. These were sites designated for de-Nazification, but in reality any critics of the new regime were imprisoned, often arbitrarily and without trial. As Bautzen I still serves as a prison to this day, an exhibition about this period could not be installed and was, therefore, relocated to Bautzen II.

Bautzen II was built in 1906 as a court jail for the Saxon Ministry of Justice. At this time, the prison was used for short gaol sentences for minor offenses (Sächsische Gedenkstätten 2016). From 1924 onwards, Bautzen II was part of the Bautzen State Penitentiary (Bautzen I) as a pre-trial confinement facility. Under the Nazi regime from 1933 to 1945, Bautzen I remained officially a prison, but Bautzen II was then transformed into a so-called 'protective prison'; people who questioned the Nazi regime were interrogated onsite and later sent to concentration camps.

After the end of the Second World War, Bautzen II was taken over by the Soviet Secret Police from 1945 to 1949, imprisoning people who disobeyed the communist system. Under inhumane conditions, prisoners were often pressured into signing fabricated confessions. In September 1949, the Soviet Secret Police transferred Bautzen II to Saxony's Ministry of Justice and once again it became a penal institution as Bautzen I was now used as a Soviet Special Camp. In 1951, the interior ministry of the GDR took over Bautzen II's management as a satellite prison to Bautzen I, before in 1956 the GDR's Ministry for State Security (MfS, also known as the Stasi) announced that Bautzen II would be a high security prison for opponents of the regime; it included prisoners from West Germany, spies and other prisoners with a special status. Bautzen II's history as a notorious Stasi prison had begun (Fricke and Klewin 2007).

3.4.1 The history of the Soviet Special Camp 1945–1950

Soviet Special Camps were introduced in the SBZ (Soviet Occupation Zone) in 1945 by the Soviet Secret Police (Greiner 2009). In total, 154,000 Germans were kept in these special camps under inhumane conditions. Many of those imprisoned died from malnutrition and disease. Although Stalin ordered the final closure of the Soviet Special Camps in 1950, most prisoners were not released; they were deported to the USSR, transferred to other prisons in the GDR to complete their sentences, or subject to arbitrary trials in Waldheim (Saxony). Those prisoners who were freed had to sign a declaration that they would never speak about their experiences. Thus, both governments, the GDR and the USSR, denied the existence of these camps. This was a state-sanctioned silence that led to a complete forgetting of this part of German history (Connerton 2008). Hence, these camps became known as *Schweigelager* (silent camps) as prisoners did not have any contact with the outside world during their imprisonment, and on release their experiences did not fit into the narrative of the German–Soviet friendship (Saunders 2018).

After the political change in 1989, access to the prisons' files in the former Soviet Union was granted, revealing the true nature of these camps. Bautzen I started its operation as a Soviet Special Camp in May 1945. In total, 27,285 prisoners were registered between May 1945 and March 1950, among them 613 were women, a very high number (Liebold and Pampel 2009). Although Bautzen's Soviet Special Camp was established for Nazi functionaries, it very quickly changed into a prison for those persecuted by the SMT (Soviet Military Administration), often civilians who were sentenced, using unfounded accusations. The Soviet authorities did not often conduct judicial hearings and the decision of guilt was frequently based on a subjective opinion. According to current knowledge, 2,700 prisoners did not survive the camp. The most common reasons were malnutrition, tuberculosis and the loose description of 'heart failure', probably more likely to be the result of other underlying conditions.

Contact with the outside world was strictly forbidden and relatives were often not informed about the fate of the prisoners. After the closure of Bautzen I Soviet Special Camp, the newly formed East German police took over the management of the prison with 6,000 prisoners remaining onsite (*ibid*).

3.4.2 Bautzen II as a Stasi prison from 1956 to 1989

In August 1956, the Ministry for State Security in the GDR decided to establish Bautzen II as a special prison for people who posed a risk to state security (Fricke and Klewin 2007). The conditions in the prison cells in the 1950s and 1960s were poor and at times inhumane. One prison cell consisted of an earth closet, a wooden bed with some straw as a mattress and a steam pipeline as a heating system (*ibid*). Former prisoners described the smell in the summer months as unbearable. The heating only functioned twice a day in the winter, making it impossible for prisoners to stay warm. At the end of the 1960s, cells were modernised with water toilets, a new heating system was installed and furniture for the cells was purchased. Hence, prisoners from the early years have different memories compared to those from the later years.

The prisoners themselves were divided into *Kommandos* depending on their status, that is, there was a *Kommando* for foreigners and one for former army/ police staff (*ibid*). The *Kommandos* were strictly separated, so that prisoners were unable to communicate and exchange information. Bautzen II also contained isolation cells for prisoners who were regarded as a severe threat to state security. These prisoners were unable to leave the cell, often for years. The day-to-day routine of prisoners was strictly organised and focused on 'betterment'. They were required to work, which was a privilege in the 1950s, yet compulsory by the 1980s. Prisoners received a wage for their work, which they could keep and later spend in the local shop to supplement the often poor prison diet. In the early years of Bautzen II, the prisoners were allowed 30 minutes of exercise time which consisted of walking up and down the small yard. However, this changed in the 1980s and a cultural programme was introduced: prisoners were allowed to watch television from time to time and borrow books from the library.

Although physical torture was not used, for most prisoners the stay in Bautzen II had serious psychological consequences (*ibid*). Imprisonment was usually an arbitrary decision, and once they arrived at Bautzen II (which they often only realised after several months), prison officials stripped the victims of their identity by creating prisoner numbers, the only form of identification they were able to use. If, for any reason, they did not adhere to the disciplinary regulations, they could have been transferred to the so-called tiger cages or the solitary confinement cells. The tiger cages were cells within a cell: the toilet was separate and, although prisoners could see it, they could not use it, a form of psychological torture. Solitary confinement cells were located within the *Hörgang*, an area where all cells were padded, so that the prisoner could not hear any sounds. S/he was also not allowed to have any outside contact. Prolonged solitary confinement under such conditions led to severe psychological damage. And even if the prisoner escaped these cells, the hard and often mundane work for many years in the basement of Bautzen II destroyed any remaining spirit.

The emerging political change in 1989 led to a first amnesty, signed in the GDR in October 1989. Prisoners who had been imprisoned for trying to leave the GDR were released (Fricke and Klewin 2007). It had little impact on Bautzen II as only a small percentage were former escapees. After the fall of the Berlin Wall on 9 November 1989, large demonstrations continued in the GDR, amongst them a candle-lit demonstration demanding the release of prisoners in Bautzen II. Despite the political change, prisoners remained at Bautzen II (and other political prisons in the GDR), resulting in a hunger and work strike on 6 and 7 December. On 8 December 1989 a further amnesty was signed and all political prisoners were scheduled to be released (*ibid*). In Bautzen II the last political prisoners left on 22 December 1989. Bautzen II's history as a Stasi prison ended.

3.4.3 The development of Bautzen II as a memorial

After the last prisoners left in December 1989, the town council decided to reuse the site as a general prison (Rudnick 2014). In the meantime, the *Bautzen Kommitee* had formed, consisting of former prisoners who were determined to research the history of the Soviet Special Camp. The town and district councils decided to support this work by providing funding for a part-time post. The committee then developed the idea of building a documentation centre at *Karnickelberg*, the location of a suspected mass grave, which Saxony's main political party CDU supported. The CDU, however, also lobbied for a memorial at Bautzen II that incorporated the history of the Soviet Special Camps, a notion that was not embraced by the *Bautzen Kommitee*. They wanted two separate memorials: one commemorating the victims of the Soviet Special Camp and another for the victims of the Stasi prison. The visit of Helmut Kohl (the German chancellor) in 1992 was therefore a welcome opportunity to convince him of these ideas. He subsequently promised to provide his support for a memorial to the victims of the Soviet Special Camp.

Shortly after Helmut Kohl's visit, the *Bautzen Kommitee* organised a large-scale archaeological excavation at the *Karnickelberg* site to search for mass graves of Soviet Special Camp victims. In total, 176 skeletons were found and re-buried at the chapel of the St. Marien Cemetery in Bautzen. The discovery of the mass grave increased the pressure to provide a memorial. Consequently, the *Bautzen Kommitee* proposed the launch of an overall memorial foundation in addition to a research institute for totalitarian regimes at the University of Dresden, now the Hannah-Arendt-Research Institute (HAIT). However, the committee then changed its mind and subsequently supported the idea of an overall memorial at Bautzen II. A sub-committee consisting of former Stasi prisoners was formed, which developed the first exhibition concept. In October 1992 rumour had it that Bautzen II was to be transformed into a modern office block. The *Bautzen Kommitee* protested immediately, writing directly to Kurt Biedenkopf, Saxony's minister at the time (Rudnick 2014). This action resulted in the first public debate concerning the memorial, ending in the final decision of developing Bautzen II as a memorial managed by the *Stiftung Sächsische Gedenkstätten* (Saxony's Memorial Foundation). This secured the long-term financial and political support for the site. Bautzen II Stasi prison memorial was born.

Unfortunately, a visit to Bautzen II in 1993 revealed the desolate condition of the building: rooms were under water, doors broken and technical equipment had been removed (*ibid*). This was heavily criticised as a conscious destruction of historical evidence. Thus, the *Bautzen Kommitee* was asked to manage the site as a trust from then onwards and began collecting material/furniture for the new exhibition. In June 1993, the new HAIT research centre at the University of Dresden was commissioned to develop the new exhibition concept using some of the material researched by the *Bautzen Kommitee*. In July 1993, Karl Wilhelm Fricke presented the new concept, suggesting a memorial as a site of remembrance and a site for dialogue with former prisoners (Rudnick 2014). The museum was also supposed to commemorate the prisoners of the Nazi era and should therefore be a space for the remembrance of the persecution of political dissidents. Hence, Fricke added a new layer to the memorial: the Nazi era prisoner. The proposal is symptomatic of Germany's struggle with the 'double past'. Whilst there is a commitment to commemorating the victims of the communist regimes, the Nazi era plays a dominant role. Integrating these different historical narratives within confined memorial spaces has led to serious debates at Sachsenhausen and Buchenwald, and continues to haunt sites such as Torgau, a former GDR youth prison that had previously been a prison for *Wehrmacht* deserters.

At Bauzten II, acknowledging the Stasi prison, the Soviet Special Camp and the Nazi prison in the same place has also led to disputes. The PDS (the left-wing party in Germany, now called *Die Linke*) and the *Opferverbände* (victims' trusts) claimed that the comparable treatment of victims of the communist and Nazi periods wilfully undermined the importance of commemorating the Stasi victims (Rudnick 2014). These arguments also had an impact on the newly-appointed managing director of the *Sächsische Gedenkstätten*, Dr Haase, who was accused of not recognising the GDR victims sufficiently. In addition, the content of the exhibition was subject to intense arguments, for example, the reconstruction of prisoner cells and the large number of personal biographies in comparison to the rest of the exhibition was viewed as inappropriate. Some of these conflicts were resolved when it was discovered that the leader of the *Bautzen Kommitee* had been a former Stasi informant (he had also been an American spy), which led to his resignation. In 1996/1997 the historians Klewin, Liebold and Wenzel presented a new concept for the exhibition (*ibid*). They suggested a clear distinction between the different time periods and the two prisons Bautzen I and Bautzen II. In addition, the over-emphasis on the victim's perspective was 'toned down' and the 'politically loaded' interpretation favouring one perspective was removed in an attempt to neutralise the exhibition. The aim was to encourage visitors to an open dialogue and to provide a space for forming their own opinions. This concept exists to this date at the site.

In summary, the initial concept for Bautzen II as a memorial was not designed to commemorate the Stasi victims. The political change in 1989 finally provided an opportunity to lift the silence over the Soviet Special Camps and their victims. There was a strong drive to break through what Connerton (2008) calls 'repressive erasure'. This led to serious conflicts over the importance of victims, that is, which victim should be represented and in which format. A decision was eventually made to give the victims of all time periods a voice, yet the current exhibition emphasises the GDR period. At the time of my fieldwork, intense debates had commenced about the new exhibition featuring the Nazi period. It was unclear how this very different time period could be integrated into the overall exhibition design. Furthermore, a discussion had erupted on whether the memorial should exhibit Ernst Thälmann, a communist celebrated in the GDR, by using a statue that was formerly located at Bautzen's central market square. The overarching management team of Saxony's memorial trust based in Dresden felt that a communist could not be exhibited, even though he was a victim of the Nazi regime. Nevertheless, 'Ernst Thälmann' arrived during my fieldwork and was initially stored in the basement where he remained, as I discovered during my visit four years later in 2020.

In 2018, the exhibition *Haft unterm Hakenkreuz* (imprisonment under the Nazi regime) was finally opened. Located on the ground floor, opposite the former Stasi offices, it follows the concept of *exemplification*. Key examples such as biographies of prisoners, important documents and special events explain the history of Nazi imprisonment. The absence of historical sources made it impossible to exhibit a complete history of the prison system during the Nazi regime, and consequently artefacts (e.g. a document), which are representative of Nazi imprisonment, are exhibited. A key aim of the exhibition is to show how the prison was embedded in Bautzen's societal structures by displaying local biographies, for example, a Sorbian artist. The exhibition focuses on individual destinies, which follows the design of the rest of the site. Visitors gain insights into the radicalisation of Nazi imprisonment, the use of prisoners as forced labourers within the local economy and the persecution of Czech resistance fighters, for example, Julius Fučík. Exhibiting Fučík leads, however, to new memory conflicts. For the victims of the Soviet Special Camp, 'celebrating' a Czech communist (hanged in Plötzensee) constitutes a trivialisation of their suffering. A similar controversy surrounds the Communist Ernst Thälmann, who was shot in Buchenwald but was also held responsible for the radicalisation of the KPD under Stalin, which links him to Stalinist crimes. Although Thälmann has a place in the exhibition, it is accompanied by an extensive explanation of the Thälmann cult in the GDR which, according to the exhibition designers, led to an unwillingness to confront the Nazi past in the GDR.

3.4.4 *The Soviet Special Camp and the Stasi prison exhibitions*

Bautzen II views itself as a 'scientific and visitor-orientated' museum that wants to exhibit the complexity of history by focusing on the dictatorships in the 20th century, and in which visitors cannot escape the different historical layers (Herold 2020). The concept of the memorial aims to 'work against the forgetting and suppressing of the memory of political violence' where 'the examination of the past takes place in the form of an open dialogue'(Thomas 2013, 6). Therefore, visitors are encouraged to explore the site for themselves with no recommended routes. However, the memorial site offers a range of themed guided tours such as 'Soviet Special Camp', 'Stasi Prison' or 'On the track of Walter Kempowski'.

Since my research took place before the opening of the Nazi exhibition, I will focus here on a more detailed description of the Stasi prison and Soviet Special Camp exhibition. The visitor enters the site through the main gate, which was previously the entrance for the well-known East German prison vans. A prison van, for many East Germans the ultimate symbol of Stasi imprisonment, is displayed at the beginning before visitors continue to the 'cinema' where an introductory film about the two prisons in Bautzen is shown (and since 2018 with an additional explanation about the Nazi prison system). The film aims to clarify the difference between Bautzen I and Bautzen II, whilst also explaining their complex history. The visitor then continues to the main building and is immediately confronted with the prison atmosphere.

The first exhibition rooms are designed to provide visitors with an overview of the different historical layers before continuing to the main part of the prison. The ground floor also explains the Stasi and its surveillance techniques. The visitor can listen to recorded conversations between prisoners and staff, or between prisoners and their relatives. In addition, the exhibition displays the biographies of the former prison officers, two of whom are still alive, with one still living in Bautzen. The focal point of the main exhibition are the personal biographies presented next to the prison cells. These biographies often show well-known prisoners (within a German context) such as Georg Dertinger, the first foreign minister of the GDR. They are also very emotionally affecting; for instance, the story of Bodo Strehlow who had steered a naval ship into West German waters during his military service, locking his crew into one of the cabins. His escape attempt was unsuccessful, and he was sentenced to lifelong imprisonment in solitary confinement at age 20. Strehlow was released after the fall of the Berlin Wall after 10 years in solitary confinement.

In a separate section the memorial also explains the experiences of the women who were imprisoned at Bautzen II. By contrast, although Bautzen II was the only political prison that held West German prisoners once they had received a sentence, this story is neglected (apart from the biography of Sigrid Grünewald). The West Germans were accommodated on the upper floors with more 'comfortable' cells, yet they are not accessible to the public. Overall, the presentation of prison cells was a particular challenge for the management team as the conditions had changed over the years. Unlike Hohenschönhausen in Berlin, Bautzen II's management team decided to reconstruct prison cells to showcase the different time periods. Therefore, the visitor can compare a prison cell from the early years with one from the later years. Visitors can also see the solitary confinement cells, the bathrooms and the limited outdoor exercise space. One of the solitary confinement cells features a sound installation that simulates a prisoner's experience. According to scientific research, a brain requires the senses (e.g. hearing, smelling) for constant stimulation. If sounds are absent, after a while, the mind will start to play tricks and one can hear sounds that are not present. Visitors can sit in the cell with headphones that are initially silent before playing a variety of 'strange' sounds. This sort of interpretative intervention stands in stark contrast to the concentration camp memorials where the 're-creation of a prisoner's experience' is rejected.

The history of the Soviet Special Camp is explained on the second floor. Similar to the Stasi prison exhibition, it uses personal biographies of perpetrators and victims to exhibit the system of the camps. Since the Soviet Special Camp was not located at Bautzen II, the memorial team decided to reconstruct a cell to represent the conditions in the 'Yellow Misery'. Together with the exhibition, the reconstruction of the cell provides visitors with a feeling of what imprisonment under the SMT meant for the individuals, but also emphasises that Bautzen II was not the location of the Soviet Special Camps. Yet, as I will demonstrate in Chapter 4, this intention does not necessarily reach the visitor.

Due to the minimal exhibition design, the 'atmosphere' of a Stasi prison is felt throughout. This gives visitors the impression that it is still a prison rather than a memorial site. Since the key aim of the memorial site is not to overburden visitors and enable them to form their own opinions (Thomas 2013), no images of suffering are shown, and no former prisoners are used as tour guides (*ibid*). While I understand the desire not to overwhelm visitors, the management team has overlooked the impact of the prison atmosphere itself. It is evident from the reactions in the visitor book, and later from the visitor research, that emotions can run high. However, not only visitors perceive the site as challenging, the memorial site also has a complex relationship with the local community.

3.4.5 Bautzen II within the wider tourism landscape

Bautzen is located within the *Oberlausitz* region, a picturesque area 50 km east of Dresden that also functions as the 'gateway' to the Sorbian heritage. The Sorbs are an ethnic minority within Germany whose traditions and language enjoy special protections. Since the fall of the Berlin Wall, Bautzen's medieval centre and boundary wall have been carefully restored, so that Bautzen is now also widely known as the town of the 'medieval towers'. Besides its medieval heritage, Bautzen is also famous for its mustard, which is celebrated in the towns mustard museum and in various restaurants that feature local dishes. Thus, Bautzen has a range of sights that make it attractive for a thriving tourism industry, yet it is overshadowed by the narrative of the Stasi prison, much to the annoyance of the local community.

Shortly after the opening of the memorial site, Bautzen attempted to move away from the image of the 'Stasi town' by not providing any signage to the memorial. Leaflets were only available on request in the tourist information centre and the memorial was not shown on local maps, with the justification that it was too far away from the town's centre (it is a 15-minute walk). In 1998, the hotel chain Holiday Inn run a marketing campaign with the slogan 'Ab nach Bautzen' (off to Bautzen), a phrase frequently used in the GDR to describe a potential imprisonment which divided the community (Hattig 2010). Some thought it was funny, others thought it was distasteful. However, it stimulated a debate about the memorial's place within the town and proved a significant turning point. The interest in the memorial rose and it was then included in annual cultural events. However, the town is still uncomfortable with its past and often prefers vague descriptions such as 'horrors of the past'. In fact, I encountered the reluctance to provide directions to the memorial during my first visit in Bautzen 2015 when I was greeted with the phrase 'what do you want to go there for?' after asking a local citizen. And during my research in 2016, I stayed in a local residence hall whose social worker explained to me that she was exasperated with Bautzen's reputation as the location of a Stasi prison, which she encounters whenever she leaves the local area. Yet Bautzen's Stasi past is not its only challenge. During my research in September 2016, violent clashes erupted between asylum seekers and the far-right, which made the headline news in Germany and abroad. A constant police presence in the town centre was a feature of my stay in Bautzen, and I often witnessed visitors at the memorial site commenting about Bautzen's politically heated situation, which continues to this day.

3.5 Summary

German memorials are sites of remembrance, cemeteries, perpetrator sites, museums and educational places (Knoch 2018). More than anything, however, these *lieux de mémoire* (Nora 1989) are representative of the memorial culture of that time. In Flossenbürg the desire to move on led to the destruction of historical evidence. Furthermore, the focus on the German victim, the refugees from the Eastern Territories, and later on the German resistance, launched a period of selective memory. Hence, West German memory politics are imprinted onto Flossenbürg's memorial landscape.

At Ravensbrück, the GDR's narrative also left its mark. The Communist victims and, crucially, the fate of the future mothers was emphasised, whilst undesirable victim groups, for example, the 'fallen' women, were excluded. Ravensbrück was also a celebration of the bright, communist future that was yet to come. The atrocities of the fascist past caused by capitalism had, after all, finally been overcome with the launch of the 'better' state.

After reunification, Germany needed to develop a common narrative. Suddenly faced with the responsibility for the East German memorial sites, Germany created an overarching memorial concept, therefore directly impacting on the future of the memorial sites. Flossenbürg and Ravensbrück were subject to what Violi (2012) called 'pathemic conservation'. The landscape was returned to a previous point in time by reinstating camp boundaries. In the case of Ravensbrück this required the removal of the remnants of the Soviet army base. Hence, the visitor today has to negotiate different layers of memorialisation.

The House of the Wannsee Conference is an example of how Germany struggled to address the perpetrator. Initially a neglected site, it opened as a memorial site in 1992 with a focus on confronting the Germans with the atrocities. The shock pedagogy exhibition was eventually replaced by a more nuanced and more accessible one, avoiding an overburdening of the visitor. Furthermore, the 2008 exhibition was designed to avert any identification with the perpetrator. Yet the House of the Wannsee Conference faced a new challenge: with an increased number of geographically more central memorial sites in Berlin, in particular the Holocaust memorial, it needed to reposition itself, which led to a new exhibition in 2020.

By contrast, Bautzen II Stasi prison memorial is representative of Germany's difficulties in exhibiting the 'triple past'. The memorial deals with the legacies of the GDR dictatorship, yet also acknowledges the period of the Soviet Special Camp and imprisonment under the Nazi regime. Unlike the exhibitions at the concentration camp memorials or the House of the Wannsee Conference, Bautzen's approach is minimalist with a strong focus on personal biographies, and the 'recreation' of prisoners' experiences in the former solitary confinement area is used to emotionally engage visitors.

At the memorial sites that commemorate the Nazi past, the *Beutelsbacher Konsens* is strictly adhered to, thus emotional engagement is only allowed in the form of eyewitness accounts but not by reconstructing 'experiences', which explains the rather clinical approach to exhibition design (Knoch 2018). Gad Yair (2014, 492) flippantly suggests that the attitude of the German memorial sites seems to be 'this is what happened, do with it what you like', losing the visitor in the process. Indeed, the information-intensive exhibitions can leave some visitors disengaged, while at Bautzen II visitors seem unable to cope at times, which I will explore in more depth in the following chapter.

3.6 Bibliography

Beßmann, Alyn, and Insa Eschebach. 2013. *Das Frauen-Konzentrationslager Ravensbrück: Geschichte und Erinnerung. Ausstellungskatalog.* Berlin: Metropol-Verlag.

Benz, Wolfgang, Barbara Diestel, and Angelika Königseder. 2007. *Flossenbürg: Das Konzentrationslager Flossenbürg und seine Aussenlager.* München: C.H. Beck.

Bormann, Natalie. 2018. *The Ethics of Teaching at Sites of Violence and Trauma: Student Encounters with the Holocaust.* New York: Palgrave Macmillan.

Brebeck, Wulff E. 1995. 'Zur Darstellung der Täter in Ausstellungen von Gedenkstätten der Bundesrepublik: eine Skizze'. In *Praxis der Gedenkstättenpädagogik: Erfahrungen und Perspektiven*, edited by Annegret Ehmann, Wolf Kaiser, Thomas Lutz, Hanns-Fred Rathenow, Cornelia vom Stein, and Norbert H. Weber, 296–300. Opladen: Verlag für Sozialwissenschaften.

Brink, Cornelia. 1998. *Ikonen der Vernichtung: Zum öffentlichen Gebrauch von Fotografien aus nationalsozialistischen Konzentrationslagern nach 1945.* Berlin: De Gruyter.

Bruha, Toni, Maria Berner, Herma Löwenstein, Anna Poskocil, Anna Schefzik, Hermine Huber, Irma Trksak, et al. 1945. *Frauen-Konzentrationslager Ravensbrück. Geschildert von Ravensbrücker Häftlingen.* Wien: Stern-Verlag.

Connerton, Paul. 2008. 'Seven Types of Forgetting'. *Memory Studies* 1 (1): 59–71.

Dalton, Derek. 2019. *Encountering Nazi Tourism Sites.* London; New York: Routledge.

Digan, Katie. 2014. *Places of Memory: The Case of the House of the Wannsee Conference.* Basingstoke: Palgrave MacMillan.

Eschebach, Insa. 1998. 'Elemente einer nationalen und religiösen Formensprache im Gedenken'. *Zeitschrift für Religions- und Geistesgeschichte* 50 (4): 339–355.

———. 2008. *Ravensbrück. Der Zellenbau.* Berlin: Metropol-Verlag.

———. 2011. Soil, Ashes, Commemoration: Processes of Sacralization at the Former Ravensbrück Concentration Camp'. History and Memory 23 (1): 131–156.

———. 2015. 'Museale Entwicklungen in ostdeutschen KZ-Gedenkstätten vor und nach dem Fall der Mauer'. In *Entnazifizierte Zone?: Zum Umgang mit der Zeit des Nationalsozialismus in ostdeutschen Stadt- und Regionalmuseen*, edited by Museumsverband Brandenburg, 1–24. Bielefeld: Transcript Verlag.

Eschebach, Insa, and Katja Jedermann. 2007. 'Sex-Zwangsarbeit in NS-Konzentrationslagern'. *Feministische Studien* 25 (1): 122–128.

Franke, Vera. 2021. 'Alles neu am Wannsee!' In *Design für Alle. Standard? Experiment? Notwendigkeit? Das Making of zur 3. Dauerausstellung in der Gedenk- und Bildungsstätte Haus der Wannsee-Konferenz*, edited by Elke Gryglewski, Hans-Christian Jasch, and David Zolldan, 69–80. Berlin: Metropol-Verlag.

Fricke, Karl Wilhelm, and Silke Klewin. 2007. *Bautzen II: Sonderhaftanstalt unter MfS-Kontrolle, 1956 bis 1989, Bericht und Dokumentation.* Dresden: Sandstein Verlag.

Geißler, Cornelia. 2015. *Individuum und Masse – Zur Vermittlung des Holocaust in deutschen Gedenkstättenausstellungen*. Bielefeld: Transcript Verlag.

Greiner, Bettina. 2009. 'Speziallager? Was für Speziallager? Zum historischen Ort der stalinistischen Verfolgung in Deutschland'. *Mittelweg 36* (3): 93–112.

Gryglewski, Elke. 2016. Personal conversation. *Exhibition Design House of the Wannsee Conference*, 25 August 2016.

Hattig, Susanne. 2010. 'Bautzen, die Stadt der Gefängnisse. Der Umgang der Stadt mit ihrer Vergangenheit'. Informationen des Sächsischen Museumbundes e.V. 40: 77–81.

Heigl, Peter. 1989. *Konzentrationslager Flossenbürg. In Geschichte und Gegenwart. Bilder und Dokumente gegen das zweite Vergessen*. Regensburg: Mittelbayerischer Verlag.

Helwerth, Ulrike. 1990. 'Gewürdigt wurden nur die Kommunistinnen'. *Die Tageszeitung*, 26 October.

Herold, Cornelia. 2020. 'Repressionsort Bautzen. Das Ausstellungskonzept "Sowjetisches Speziallager Bautzen" und die Konzeption "Stravollzug 1933–1945" in den beiden Bautzener Haftanstalten'. In *Erinnerungs- und Gedenkorte im Sächsischen Dreiländereck Polen – Tschechien – Deutschland*, edited by Sächsische Landeszentrale für politische Bildung and Umweltbibliothek Großhennersdorf e.V., 151–171. Dresden: Sächsische Landeszentrale für politische Bildung.

Heyl, Matthias. 2016. 'Mit Überwältigendem überwältigen? Emotionalität und Kontroversität in der historisch-politischen Bildung. Ein Plädoyer für die Schärfung des Profils historischer Bildung'. In *Politische Bildung auf schwierigem Terrain: Rechtsextremismus, Gedenkstättenarbeit, DDR-Aufarbeitung und der Beutelsbacher Konsens*, edited by Jochen Schmidt, Steffen Schoon, and Landeszentrale für Politische Bildung Mecklenburg-Vorpommern, 37–55. Schwerin: Landeszentrale für Politische Bildung Mecklenburg-Vorpommern.

Jäckel, Eberhard. 1992. '"Wo Heydrich seine Ermächtigung bekanntgab" – Der Holocaust war längst im Gange'. *Die Zeit*, 17 January. www.zeit.de/1992/04/die-konferenz-am-wannsee.

Jansen, Elmar. 1959. 'Mahnmal auf blutgetränkter Erde'. *Neue Zeit*, 6 September.

Jasch, Hans-Christian. 2017. 'The House of the Wannsee Conference: Tourism and Holocaust Education at a Perpetrator Site'. *Worldwide Hospitality and Tourism Themes* 9 (2): 146–157.

Knoch, Habbo. 2018. 'Gedenkstätten'. *ZZF: Centre for Contemporary History*. https://doi.org/10.14765/zzf.dok.2.1221.v1.

Köhler, Reinhard, and Jan Thomas Plewe. 2001. *Baugeschichte Frauen-Konzentrationslager Ravensbrück*. Schriftenreihe der Stiftung Brandenburgische Gedenkstätten. Berlin: Edition Hentrich.

Kühling, Gerd. 2008. 'Schullandheim oder Forschungsstätte? Die Auseinandersetzung um ein Dokumentationszentrum im Haus der Wannsee-Konferenz (1966/67)'. *Zeithistorische Forschungen* 5 (1): 211–235.

Kühling, Gerd, and Hans-Christian Jasch. 2017. 'Wer hier weint, hört nicht mehr auf. Zum Umgang mit der Wannsee-Konferenz und ihrem historischen Ort'. *Zeitgeschichte Online*. https://zeitgeschichte-online.de/kommentar/wer-hier-weint-hoert-nicht-mehr-auf.

Lammert, Marlies. 1965. *Will Lammert: Ravensbrück*. Berlin: Deutsche Akademie der Künste zu Berlin.

Liebold, Cornelia, and Bert Pampel. 2009. *Hunger – Kälte – Isolation: Erlebnisberichte und Forschungsergebnisse zum sowjetischen Speziallager Bautzen 1945–1950*. 5th ed. Dresden: Stiftung Sächsische Gedenkstätten.

Litschke, Egon. 1985. 'Museum des Antifaschistischen Widerstandskampfes in der Nationalen Mahn- und Gedenkstätte Ravensbrück'. *Neue Museumskunde* 85 (2): 115–119.

Meier, Günter. 1960. 'Über die künstlerische Gestaltung des Lagermuseums in Ravensbrück'. *Bildende Kunst* 2: 90–93.

Ministerium für Wissenschaft und Kultur Brandenburg. 2009. 'Geschichte vor Ort: Erinnerungskultur im Land Brandenburg für die Zeit von 1933 bis 1990'. https://mwfk. brandenburg.de/media_fast/4055/Konzept_GeschichtevorOrt.pdf.

Möller, Lena. 2019. *'Auf Stätten des Leids Heime des Glücks': Die Siedlung am Vogelherd auf dem Areal des ehemaligen KZ Flossenbürg und ihre Emotionalisierung als Wohn- und Gedächtnisort.* Münster: Waxmann Verlag GmbH.

Nora, Pierre. 1989. 'Between Memory and History: Les Lieux de Mémoire'. *Representations* (26) (April): 7–24. https://doi.org/10.2307/2928520.

Oswalt, Philipp, and Stephanie Oswalt. 2000. 'Entwurf zur Gestaltung der erweiterten Gedenkstätte Ravensbrück'. In *Das Mädchenkonzentrationslager Uckermark*, edited by Katja Limbächer, Maike Merten, and Bettina Pfefferle, 280–292. Münster: Unrast Verlag.

Oswalt, Philipp, and Stefan Tischer. 1998. 'Ehemaliges Frauen-KZ Ravensbrück'. Unpublished.

Paver, Chloe E.M. 2018. *Exhibiting the Nazi Past: Museum Objects between the Material and the Immaterial.* Cham: Springer.

Pearce, Caroline. 2011. 'Visualising "Everyday" Evil: The Representation of Nazi Perpetrators in German Memorial Sites'. *Holocaust Studies* 17 (2–3): 233–260.

Peters, Ursula. 2015. 'Zu Will Lammerts Figurenentwürfen für das Mahnmal im ehemaligen Frauenkonzentrationslager Ravensbrück, 1957'. *KulturGut: Aus der Forschung des Germanischen Nationalmuseums* (40): 19–24.

Porsdorf, Friedrich. 2019. Personal communication. *Re-design of the Exhibition at Ravensbrück in 1980*, 1 July 2019.

Rapson, Jessica. 2015. *Topographies of Suffering: Buchenwald, Babi Yar, Lidice.* New York: Berghahn Books.

Roseman, Mark. 2002. *The Villa, the Lake, the Meeting: Wannsee and the Final Solution.* London: Penguin Press.

Rudnick, Carola S. 2014. *Die andere Hälfte der Erinnerung: Die DDR in der deutschen Geschichtspolitik nach 1989.* Bielefeld: Transcript Verlag.

Rydén, Johanna Bergqvist. 2018. 'When Bereaved of Everything: Objects from the Concentration Camp of Ravensbrück as Expressions of Resistance, Memory, and Identity'. *International Journal of Historical Archaeology* 22 (3): 511–530.

Sächsische Gedenkstätten. 2016. 'Bautzen II Der "Stasi-Knast" | Gedenkstätte Bautzen | Stiftung Sächsische Gedenkstätten'. www.stsg.de/cms/bautzen/geschichte/bautzen_ii.

Saunders, Anna. 2018. *Memorializing the GDR: Monuments and Memory after 1989.* New York: Berghahn Books.

Schwarz, Erika, and Simone Steppan. 1999. 'Die Entstehung der Nationalen Mahn- und Gedenkstätte Ravensbrück, 1945–1959'. In *Die Sprache des Gedenkens. Zur Geschichte der Gedenkstätte Ravensbrück 1945–1995*, edited by Insa Eschebach, Sigrid Jacobeit, and Susann Lanwerd, 218 – 238, Berlin: Hentrich Edition.

Skriebeleit, Jörg. 2009. *Erinnerungsort Flossenbürg: Akteure, Zäsuren, Geschichtsbilder.* Göttingen: Wallstein Verlag.

———. 2011. 'Nachwirkungen eines Konzentrationslagers'. *Gedenkstättenforum – Rundbrief* 159: 21–27.

———. 2016. 'Relikte, Sinnstiftungen und Memoriale Blueprints'. In *Von Mahnstätten über Zeithistorische Museen zu Orten des Massentourismus: Gedenkstätten an Orten von NS-Verbrechen in Polen und Deutschland*, edited by Enrico Heitzer, Günter Morsch, Robert Traba, and Katarzyna Woniak, 48–65. Berlin: Metropol-Verlag.

Später, Jörg. 2013. 'Klaus Kempter: Joseph Wulf: Die Massenmörder gehen frei herum, haben ihr Häuschen und züchten Blumen'. *Sec. Feuilleton*. www.faz.net/1.2149976.

Starke, Katrin. 2017. 'Frauen-KZ Ravensbrück: Erinnerung an die "Zone des Elends"'. *Berliner Morgenpost*, 8 February. www.morgenpost.de/brandenburg/article209544059/Frauen-KZ-Ravensbrueck-Erinnerung-an-die-Zone-des-Elends.html.

Stier, Oren Baruch. 2015. *Holocaust Icons: Symbolizing the Shoah in History and Memory*. New Brunswick, NJ: Rutgers University Press.

Thomas, Marcel. 2013. 'Coming to Terms with the Stasi: History and Memory in the Bautzen Memorial'. *European Review of History: Revue Européenne d'histoire* 20 (4): 697–716.

Tischer, Stefan. 2020. Personal conversation. *Landscape Design Ravensbrück*, 10 January 2020.

Unknown. 1982. 'Rekonstruktion und Gestaltung der Mahn- und Gedenkstätte Ravensbrück'. 530 SED BL Pdm 7312. Brandenburgisches Landeshauptarchiv.

Violi, Patrizia. 2012. 'Trauma Site Museums and Politics of Memory: Tuol Sleng, Villa Grimaldi and the Bologna Ustica Museum'. *Theory, Culture & Society* 29 (1): 36–75.

Yair, Gad. 2014. 'Neutrality, Objectivity, and Dissociation: Cultural Trauma and Educational Messages in German Holocaust Memorial Sites and Documentation Centers'. *Holocaust and Genocide Studies* 28 (3): 482–509.

4 Visitor experiences at German memorial sites

Insa Eschebach (2016) argues that one of the greatest challenges for the future of the memorial sites is to research visitors. How does one conduct such research if most traditional research methods are rejected (e.g. Gudehus 2004)? Indeed, it was the dilemma that occupied my mind throughout the research process and still does as I am writing this book. Inspired by Joy Sather-Wagstaff's (2016) work on tourist experiences at the 9/11 memorial in New York, I turned to anthropological methodologies with the aim of developing a deeper understanding of tourists at German memorial sites. Since ethnographic research is sometimes criticised for being too subjective, I also conducted exit interviews using a pre-designed questionnaire.

As common in social sciences research, I conducted several pilot studies at the Holocaust Centre Nottingham, the National Memorial Arboretum Staffordshire and finally the Tyne Cot Cemetery in Belgium, the largest Commonwealth war cemetery in the world. All three sites confirmed what other researchers and memorial managers have already aptly observed: a visitor survey is not enough to capture the unique experiences at the sites. Whilst open visitor observations filled this gap, it made me acutely aware that I was intruding on often very personal experiences. A couple at the National Memorial Arboretum Staffordshire wanted to commemorate a fellow soldier who died in Afghanistan, while a visitor at the Tyne Cot Cemetery was overcome by emotions when he saw his uncle's name on the list of the missing. Hence, observing visitors at memorial sites also poses significant ethical dilemmas. How close does one get to the visitor? As Francisco Ferrándiz (2020) points out, one of the greatest challenges when conducting participant observations is to establish a rapport with the subject studied in order to allow an empathetic and open exchange. In the context of memorial visitors this is often difficult to achieve within the limited time frame visitors spend onsite. Thus, as in the 'real world', with some visitors trust could be established very quickly, with others it took a longer period of time or it was not achieved at all.

Moreover, whilst I could be viewed as a 'native' since I studied visitors in my home country Germany, the diversity of visitors required me frequently to adapt my position as a researcher. At the Stasi prison Bautzen II, my East German background assisted me greatly in creating frank conversations with

DOI: 10.4324/9781003126836-4

Bautzen's visitors, while at the House of the Wannsee Conference my German background could be seen as a hindrance with Jewish visitors, for whom the villa stands for Nazi Germany's painful planning of the murder of the European Jews. Thus, my presence and identity has certainly influenced the visit, and the results presented here could be viewed as biased. However, themes that emerged during the exit interviews often recurred during the observations, and vice versa. Furthermore, my presence at the memorial sites, and my interaction with members of staff and local residents, provided me with the additional background information and a 'feeling for the site' that allowed me to put the visitor observations into context.

Although I interviewed in total 400 visitors and accompanied in total 100 visitors on their visits, the visitors' experiences presented here cannot be a complete picture of the multitude of visits that happen every day at memorial sites. My research was conducted in the summer months between June and September 2016, and I am convinced that similar research in the depth of winter would have resulted in different outcomes. The political atmosphere – the recent Brexit referendum, Trump's election in the United States and the recent refugee crisis – influenced the topics visitors talked about when they were confronted with this violent past. Hence, under new circumstances visitors might react differently. I am also acutely aware that my interpretation of visitors' behaviour is influenced by my own upbringing, and another researcher might arrive at different conclusions. My aim for this research, however, was to provide a more nuanced understanding of tourism at memorial sites that stretches beyond the notion of the shallow, 'dark tourist' who travels to memorial sites to satisfy his/her morbid desire. In so doing, the book aims to challenge existing tourism discourses and open up new avenues for debate. In addition, the results question the often celebrated Holocaust memory culture as a form of vaccination against future genocides. Whilst I do not subscribe to the negative views that tourism adds to a trivialisation or even Holocaust denial (Cole 1999), I am not convinced that visits to memorial sites will transform visitors as defenders of democracy, a view that is widely held in many educational settings.

4.1 Visitor experiences at the concentration camp memorials Flossenbürg and Ravensbrück

Visitor research at Flossenbürg and Ravensbrück comprised 100 exit interviews as visitors were leaving the site and 25 visitor observations. Visitors were approached randomly and the sample can be said to reflect the overall visitor profile at the sites during my research. At Flossenbürg, survey participants were often in the age group 41 to 50 and 51 to 60. However, there was also a significant number of young visitors in the age group 21 to 30. Most visitors were from Germany, followed by the United States, often from the American army base nearby. Flossenbürg also had a high number of visitors with a religious affiliation (Catholicism and Protestantism), and most visitors stated *Realschulabschluss* (school certificate after 10 years) as their highest educational qualification. At Ravensbrück, the majority of visitors surveyed during the period of July 2016 were in the age group 41 to 50 and 51 to 60. They predominantly arrived from Germany, in particular from Berlin. Similar to Flossenbürg, most visitors' highest qualifications were *Realschulabschluss* or *Abitur*, yet, unlike Flossenbürg, a large proportion of visitors (46 per cent) stated that they did not belong to a religious affiliation. The visitor demographics of the ethnographic research (participant observations) varied significantly from site to site. At Ravensbrück, of the total number of visitors (n=25) most visitors were German (20) in addition to visitors from Switzerland, Poland, Britain, Belgium and the United States. Equally, at Flossenbürg, of the total number of visitors (n=25), most visitors were from Germany (22), followed by visitors from the United States, Belgium and Sweden. At Flossenbürg, most observees were in the age groups of 20 to 30 and 60 to 70 years, although the age groups 30 to 40 and 50 to 60 years were also represented. By contrast, at Ravensbrück, the most common age groups were 50 to 60 and over 70 years, with only two visitors in the 20 to 30 age group.

Flossenbürg had an unusually high number of locals from the federal state of Bavaria and repeat visitors, with 30 per cent of visitors reporting that it was not their first visit. When asked how often they had visited before, they frequently stated two or more visits. The curtain of silence hanging over Flossenbürg described in Chapter 3 appeared to encourage visitors to now find out the truth for themselves and/or engage with a legacy that was previously hidden from their view. This was also reflected in the reasons for visiting Flossenbürg, with visitors mentioning wanting to 'learn about the history' followed by 'the motorway sign', 'day out' or a 'taboo topic in school'. The motorway sign is located on the A93, a motorway built after German reunification, connecting Bavaria with the Eastern state of Saxony. Hence, this motorway is frequently used for travelling between the South and the East of Germany, and visitors often stated that 'they always wondered what Flossenbürg was'.

In contrast to Flossenbürg, Ravensbrück visitors were often from further afield and stumbled across the site accidentally while on holiday in the local area. In fact, most visitors from the former West Germany had never heard of Ravensbrück and/or had a vague knowledge about a women's only concentration camp on German soil. Consequently, when visitors were asked about their knowledge about the site prior to the visit, it was limited to 'concentration camp for women' without any further details. This stands in contrast to Flossenbürg, whose visitors could refer to specific historical aspects (e.g. the theologian Dietrich Bonhoeffer or the mass grave 'ash pyramid').

These findings point to several new insights. Firstly, the visits to these memorial sites are not carefully planned, they occur spontaneously. Hence, Germany's desire that visits to memorial sites should be accompanied by a careful before and after discussion, often stressed for school visits, fails in respect of the individual visitors. Secondly, memory narratives linger long after they have been re-written. West Germany's focus on the resistance explains why visitors at Flossenbürg still remember Bonhoeffer who was involved in the attempted assassination of Hitler in July 1944. Similarly, that many visitors from the former West Germany did not possess much knowledge about a women's concentration camp is unsurprising, given that the site was located behind the Iron Curtain and out of reach for most visitors. Thirdly, the GDR's programme of compulsory visits to the concentration camp memorials for annual anniversaries, school visits and for any other official functions clearly had an impact on the local community who, unlike at Flossenbürg, barely featured during my research. Thus, when analysing tourism to memorial sites we cannot ignore the 'host community'. Moreover, Ravensbrück and Flossenbürg are not sites that are visited predominantly by higher education graduates, who are normally over-represented amongst visitors to museums and heritage sites.

4.1.1 Impact of the exhibitions

Visitors' engagement with the exhibitions was tested by using a variety of opinion statements in the visitor survey and by accompanying visitors throughout their visits. At Flossenbürg most visitors wanted to learn more about its history, but 14 per cent also stated that they had always done it, a lower number compared to Ravensbrück but still significant. Most visitors agreed with the statement 'The visit provided me with greater insight into the concentration camp system' (23 per cent strongly agreed and 52 per cent agreed); so a visit to Flossenbürg appeared to have increased the understanding of the concentration camp system. The statement 'Did the visit question my own behaviour?' was designed to analyse whether visitors critically reflect on their visit and potentially make a connection to their own personal lives. Visitors predominantly disagreed with this statement or were unsure (15 per cent unsure, 39 per cent disagreed and 6 per cent strongly disagreed). By contrast, the statement 'the visit allowed me to work through my own experiences' seemed to divide opinions with 45 per cent agreeing with the statement, 29 per cent disagreeing and 14 per cent being unsure. A cross comparison between age groups and a response to this statement revealed that it is mostly the age group above 41 to 50 years that responds positively to this statement. These are the people who either were children during the Nazi period or grew up in the shadow of the Second World War. Although they were aware of the Second World War, they were often confronted with the wall of silence and were unable to process the experiences. Hence, a visit to Flossenbürg encourages them to deal with their own personal legacies. In addition, visitors seemed keen to continue visiting memorial sites, with 53 per cent naming Auschwitz and Dachau as the most preferred future destinations.

At Ravensbrück, most visitors agreed or strongly agreed with the statement that the visit had encouraged them to continuously engage with this particular part of German history, while almost the same number of visitors suggested that they 'had always done it', irrespective of the visit. It is a paradox given the lack of knowledge about the site prior to the visit. Thus, visitors are either over-confident or saturated from the frequent confrontation with the Nazi past which therefore leads to a lack of further engagement. On the other hand, the questions about a greater insight into the concentration camp systems and potential similar visits met mostly with positive responses. Whilst visitors do not want to engage further with historical facts, they are not averse to visiting other similar memorial sites, with Auschwitz and Buchenwald the most mentioned sites. Ravensbrück's visitors appeared divided over the question of whether the visit made them question their own behaviour, since over 40 per cent disagreed but over 30 per cent agreed or strongly agreed with the statement. Indeed, visitors were often quick to add that 'I have not lived during this period'. On the other hand, over 40 per cent stated that the visit encouraged them to work through their own experiences, but a significant number also disagreed or could not answer this question.

The most striking result is that when visitors suggest that these visits do not make them question their own behaviour they clearly distance themselves from these events. In fact, I met only one Australian couple at Flossenbürg who reflected on the treatment of Australian aborigines when asked this question, and one Swiss visitor at Ravensbrück who responded that 'this is the most crucial question'. In contrast, I encountered visitors who questioned Germany's refugee policies, arguing that they should not be 'invited'. As discussed in Chapter 2, Germany's memory culture views the Holocaust as a singular event, which discourages any comparisons with contemporary political challenges. In many ways, therefore, it is not a surprising result. Nevertheless, if Germany's desire is to ensure 'never again', then these results are dispiriting. Most German memorial managers would agree that an outcome of the visit should be the critical understanding of democracy and human rights which forms the basis for most education programmes. Yet with adult tourists who are not part of formal education programmes, this aim seems to be missed.

However, the questions 'which aspects of the site do you remember most?' and 'were there any aspects that you found particularly distressing?' as well as the close observation of visitors provide us with a more complex insight into the impact of the sites. At Flossenbürg, visitors mostly remembered the valley of death, the crematorium and the exhibition after their visit. They also declared the crematorium, the valley of death, the shower room and the death marches to be the sites and/or images that caused emotional distress. When asked which objects or images represented death or suffering, the most frequent themes were 'bodies piles up', 'death marches', 'quarry' and 'liberation'. Mieke Bal (2010) argues that to understand visitors' museum experiences one has to understand the narratology of a museum. The way an exhibition is designed influences the visitor's journey through the space: it can speed up the journey, slow it down and/or can change the direction of focus. The management team at Flossenbürg strategically placed the video footage of the 1945 death marches at the end of the exhibition in a location that is unavoidable for the visitor. In so doing, the aim was to highlight that local people watched these atrocities and failed to act, thus touching the visitor's conscience just before s/he leaves the space. Since most visitors remember the images of 'bodies piled up', this form of design appears to be successful in achieving the management team's aims.

Yet the landscape's architect vision of an experiential path with a descent into hell (crematorium) and an ascent to salvation (chapel), described in Chapter 3, also left an impression upon visitors. In fact, they frequently described the 'dark atmosphere' of the valley of death. The crematorium is, however, also part of the universal symbolism of the Holocaust (Stier 2015). These symbols stand for the mass murder and are constantly repeated in the media, in the literature and increasingly on social media. For visitors, encountering these icons is a form of 'seeing is believing', but also a way of 'experiencing' authenticity, since visitors project their own emotions onto these buildings. On observing two visitors photographing each other in Buchenwald's crematorium, Charles Maier (2002) wondered whether it is the carefully curated exhibition which remains in visitors' memory, or rather the crematorium in which one can feel the presence of the dead. This research suggests the latter, which might be disturbing for many Holocaust scholars and/or memorial managers. Indeed, Maier (*ibid*) is simultaneously fascinated and disturbed by the two tourists, but concluded that it is the aura of the crematorium that makes these locations so powerful.

It is, however, not just the aura or the exhibitions that shape tourists' perceptions of the site; the structure of the buildings themselves can also influence visitors' engagement with objects. Whilst Chloe Paver (2018) emphasises Ravensbrück's attempt to focus on intimate objects that signify women's daily life in the camp, the sterile atmosphere of the SS-*Kommandantur* restricts the visitor's engagement with these objects. Indeed, the exhibition is spread over 15 rooms, forcing the visitor to walk in and out of the rooms, therefore interrupting the natural flow. Moreover, Ravensbrück's exhibition often embeds the objects within an extensive textual content. Sandra Dudley (2010) argues that this approach can take visitors away from the materiality, that is, the object becomes insignificant without the textual context. This conventional form of museum design can inhibit a visitor's physical and emotional engagement with an object, and by extension might influence the transformative power museums can have, an issue I observed at Ravensbrück with visitors frequently commenting on feeling overwhelmed by the amount of information. When asked what they remembered most after their visit they rarely mentioned specific objects; instead they recalled the crematorium, the cell block (GDR exhibition), the execution path and the empty *Lagerplatz*. As the most distressing aspects of the site, the crematorium, the ovens, the execution path and the medical experiments were mentioned. Andrea Witcomb (2010) explains that museum objects carry a meaning and can therefore create an effect, which was the GDR exhibition's aim in the former cell block. Thus, emotional personal testimonies, harrowing images of murder and/or artistic expression of female suffering were used to engender visitors' emotions, using very little textual information. Ironically, it is this exhibition design that seems to create a lasting impression. But the Ravensbrück research results also exposed the 'gendered gaze'. It was mainly women who mentioned the experiences of children when asked what they remembered about their visit. Hence, visitors are drawn to objects or aspects of the site that resonate with them. At Flossenbürg, for instance, older German visitors were often captivated by the exhibition 'What remains', which details Flossenbürg's and Germany's struggle to come to terms with the Nazi past. American visitors, however, barely engaged with this exhibition. In essence, the impact of the exhibitions depends on their ability to evoke emotions, which in turn depends on visitors' personal relationship to the site.

4.1.2 Perception of authenticity

One of the most controversial, yet also most revealing question of the visitor survey was 'does the memorial site represent the true atmosphere of a former concentration camp?' Of course, I am aware that one cannot (and hopefully nobody would want to) represent the 'true' atmosphere of a concentration camp. The question was designed to capture visitors' perception of authenticity. Overwhelmingly visitors disagreed, stating that the inappropriate housing development at Flossenbürg, a supermarket next to the memorial site, the beautiful parkland, the lake at Ravensbrück and/or the missing barracks as reasons for the absence of authenticity. In fact, one visitor at Ravensbrück commented that there were birds singing at the memorial site, which did not occur during his visit at the nearby Sachsenhausen Concentration Camp memorial. Visitors rarely acknowledged that one will never be able to exhibit the 'true atmosphere', and those visitors who responded positively to the question argued that the 'site had a cold and dark atmosphere'.

Authenticity is a constant debate in tourism that often centres around the originality of objects. Tim Cole (1999), for instance, criticises the reconstruction of original buildings (e.g. gas chambers), which the tourist is unable to interpret, leading to the development of 'Auschwitzland'. Such discussions miss a crucial point: authenticity is a social construction and not dependent on the originality of objects. In fact, Erik Schilling (2020) suggests that authenticity is the correspondence of an observation with an expectation of the observer. In order words, our expectation needs to match what we see in front of us. When visitors note the beauty of the site, then their expectation of what an authentic concentration camp memorial should look like does not match. This 'imagined perception' is increasingly influenced by media presentations and/or prior visits to other memorial sites. Consequently, visitors frequently highlighted that 'original structures' at Auschwitz were more impressive.

Flossenbürg and Ravensbrück also reveal the importance of the emotional connections to landscapes. While human geography has long understood that places are complex constructions with physical, emotional, social and cultural features (Rickly-Boyd 2013) in memory and heritage studies, the importance of landscape design for visitors is neglected. Ravensbrück and the nearby town of Fürstenberg were predominantly known for spa facilities in the 19th century and often used by wealthy Berliners. With the construction of the concentration camp, the character of the landscape changed from freedom and relaxation to incarceration, pain and death. With the establishment of the Soviet Army base another layer was added: oppression and secrecy. The remainder of the landscape was sacralised as a memorial (Eschebach 2011). After reunification, the site's sacralisation was expanded and efforts to re-create leisure facilities were thwarted. Tourists often demanded a landing stage for their canoes at the bottom of the statue *The Burdened Woman;* it was rejected by survivors and the memorial management team as a 'profanation of the site'. As a consequence, Lake Schwedt was divided by buoys into a northern and southern part. The northern part, viewed as a grave, is strictly off-limit for water sports, while the southern part is used for leisure activities. For example, I witnessed an incident where tourists were reprimanded by the water police for leaving their canoes next to the statue. Their attempt to visit the memorial was subsequently cut short. This example highlights that a landscape has different meanings: for the survivors it is a grave, for memorial visitors it is a site of atrocities, for the local community it is a daily painful reminder of the Nazi and GDR regimes, and for water sports tourists it is a leisure facility. Tourists interact with the features of a landscape, hence the atmosphere of a place contributes to notions of authenticity. A landscape that is 'alive' will not fit into the highly moralised Holocaust memory discourse. Only a 'dead' site will honour the victims, which Ravensbrück's black clinker surface ironically reinforces. Thus, when analysing visitors to memorial sites, a sole focus on the interaction between the visitor and the museum space is not sufficient. It is the emotional connection to the site, evoked by physical traces, that will shape the concept of authenticity.

At Flossenbürg, a frequent visitor comment was 'I have no idea how anyone can live here', referring to the housing development on the foundation of the barracks. For visitors, the authenticity of the site is compromised by signs of life at the former concentration camp. The German flag, displayed at some of the houses during the European Football Championship 2016, caused outrage. Yet for the residents the houses have a different meaning. After the painful loss of their *Heimat* in the Eastern territories, the new housing development meant that they finally could call Flossenbürg 'home' again. They now observe sceptically the development of the memorial site, and often feel misrepresented in the exhibitions at the memorial site (Möller 2019). Jody Manning (2010) observed similar issues at Auschwitz where visitors expressed unease about Oświęcim's residents using former Nazi buildings as recreational spaces. David Duindam (2017) whose research at Amsterdam's Schouwburg also revealed conflicts with the local neighbourhood explains that memory cannot be easily contained within the memorial complex; they spill over to the surrounding area, particularly if it involves potential collaboration. In fact, one visitor explained during an interview that Flossenbürg's residents were often called 'KZler', a word which used to describe concentration camp prisoners, but, in this case, a derogatory term, linking the residents to the concentration camp. The main aim of the architect's first memorial in the 'valley of the death' was to use the natural boundary of the valley to contain the pain. Yet the reinstatement of the former concentration camp outline dissolved the boundary, providing visitors with ample opportunities to project their memories onto the site, which can be the subject of conflicts between the memorial site and the local community.

4.1.3 Emotions

'The confrontation with the Holocaust is highly influenced by emotions, yet it is not clear how these emotions develop and what quality they have in people who have no direct connection to this history'(Assmann and Brauer 2011, 73). There is, indeed, little understanding of how visits to memorial sites evoke emotions and how visitors consequently deal with them. The visitor survey included two questions about the emotional state before and after the visit. At Flossenbürg, the question about emotions after the visit revealed a shift from feeling 'normal' (the most commonly word used) to visitors reporting feelings of empathy, sadness, shock, worry and contemplation. This also correlates with the answers given when visitors were asked to summarise their visit in three words, with the most common phrases used 'harrowing', 'thought-provoking' and 'depressing'. At Ravensbrück, results of emotions after the visit were very similar to Flossenbürg, with the most commonly used emotions listed after the visit being empathetic, shocked, contemplative and worrying. These answers raise many questions. Visitors seemed disappointed about the lack of 'authentic' evidence, often describing the sites as 'too nice', and yet they feel 'sad' or 'shocked'.

Ute Frevert (2013) has shown in her work on the history of emotions that emotions are not only biologically acquired but also shaped culturally. Empathy, for instance, now commonly used in Western societies would have been barely known in the 19th century. Our ancestors would have referred to *Mitfühlung* (sympathy) in order to describe an understanding of another person's difficulty. Considering Germany's moral demand to acknowledge the Nazi past and to 'never forget', there is a cultural expectation to feel 'sad' after a visit. Thus, the words used to describe the visit tell us perhaps more about Holocaust memory culture, and in particular German memory culture, than what visitors truly feel. Indeed, the close observation of visitors provided a greater insight into the complexity of emotions at Flossenbürg and Ravensbrück.

The Flossenbürg exhibitions caused a range of emotions, and at times very conflicting. A family from Saxony (parents and daughter) spent a considerable time in both exhibitions. They were surprised by the items in the 'What remains' exhibition, which features a local restaurant owner who said in 1992 'that they [the prisoners] had a good life in the camp', which, according to the man of the group 'was unbelievable that such comments were made as late as 1992'. In the main exhibition both parents referred to relatives who had died in Buchenwald and whom they knew very little about. Indeed, the woman commented: 'Mum never talked about her uncle who was in Buchenwald, it's a shame really.' The daughter was particularly moved by the story of a young boy who had lost all his teeth due to malnutrition, and a girl who died shortly after liberation in a failed attempt to save her. As they subsequently explored the rest of Flossenbürg, the daughter entered the chapel for a prayer with tears in her eyes. On leaving the site, the woman summarised:

> It [such a visit] is always emotional. One can never forget it and yet the far-right scene is rising. My father has always said humanity is capable of repeating this history, there won't be a World War III but civil wars.

Another German couple was overwhelmed by the 'What remains' exhibition, and subsequently decided not to visit the rest of the site. Towards the end of the exhibition the woman commented: 'I can't cope with it anymore, I also know a lot of it from books.' On leaving the site, her partner said: 'I am unable to deal with this history. I cannot understand how people can treat other people in such a brutal way.' This, however, stood in contrast to the woman's emotions: 'The visit did not cause an emotional reaction but I could imagine that someone who has never seen it before, might be affected by it.' Hence, emotions experienced onsite are also influenced by foreknowledge. For some visitors, extensive knowledge about the Nazi past might cause a 'numbing' effect while others will use this knowledge to step into the shoes of the prisoners. Philip Meinhold (2015) demonstrates in his book *Erben der Erinnerung: Ein Familienausflug nach Auschwitz* (Heirs to Memory: A Family Trip to Auschwitz) how different family members experience a trip to Auschwitz. Being the descendant of German Jews, his family history includes people who survived Auschwitz, and yet he did not feel emotional during his trip to Auschwitz. His brother eventually left the guided tour as he could no longer absorb the facts and wanted some space to just reflect, while his sister cried throughout the tour. Meinhold also wonders whether the frequent confrontation with the Nazi past might led him to become unemotional or whether the expectations of a visit to Auschwitz are simply too high. Thus, emotions at memorial sites are very complex and unpredictable.

One woman in her 50s from Dortmund was emotionally distressed by the crematorium at Flossenbürg. She was shaking as she entered it and left seconds after. When she encountered the former location of the camp's brothel, she remarked: 'I have no idea how prisoners could exploit other prisoners', at which point her partner replied in a rather cynical way: 'Carrot and stick, keep them motivated'. As they were leaving, the woman summarised the visit in tears: 'This is so awful, I hope I'll never have to experience something like that.' According to her partner, this was the first time they had visited a concentration camp memorial. Similarly, a couple from Bavaria was distressed by the images of the liberation shown in the exhibition, and as soon as corpses were shown, the male visitor left the exhibition. The 'valley of death' engrossed them, though they walked hesitantly into the crematorium and left it very quickly. The man told me that his father was a prisoner of war in Russia, returning home disabled. His father never talked about his experience, so he did not know what his father had witnessed. Both felt that they were emotionally overwhelmed by Flossenbürg's main exhibition.

One American family consisting of a grandfather, daughter and grandson displayed a fascinating generational dynamic. The main driving force was the grandson's interest in history and geocaching (a modern version of a scavenger hunt). Whilst the grandson engaged extensively with the 'historical clues' around the site, his mother struggled with some of the imagery; in particular the images of prisoners being hanged caused emotional distress and she was hesitant about entering the crematorium. Furthermore, the American memorial plaque caused a period of reflection where she commented: 'When I think about how old these stones are and how many people were here.' She also complained about the behaviour of German schoolchildren who were laughing and joking, stating that this was inappropriate behaviour. One could, of course, argue that geocaching is not a suitable activity for a concentration camp memorial either, with which I would disagree having observed the family. Looking for clues meant that they extensively engaged with the wider memorial landscape which most visitors would not even notice. Moreover, the geocaching software (not provided by the memorial) provided them with additional historical information at each stop and led to conversations between the family members. Thus, geocaching was an educational family activity which the German memorial sites currently do not provide since their focus is on the school visitors. At the end of the visit, the woman explained that these memorials are important for her as her husband was fighting in Afghanistan. In addition, her own great grandfather had fought and died during the First World War in France, where she had visited and described as very emotional. For this family, the visit to Flossenbürg was a personal connection to their own past and present military history.

Projecting emotions onto physical features was also a key feature at Ravensbrück. One German couple commented that 'the youth hostel was macabre and there must be a negative atmosphere from these buildings'. Similarly, a couple who had stayed overnight concluded that 'they slept well the first night, but not anymore afterwards'. Another couple I accompanied overheard a tour guide who explained that the GDR had flattened the ashes to build the new memorial and that sometimes ashes of prisoners were thrown into the lake. The woman reacted horrified: 'Unbelievable, now I'm even walking on the ashes.' At the end of the visit, the woman commented: 'the ovens are the worst, that's enough now, I've had enough'. Nevertheless, she also seemed to experience an internal conflict:

> I'm annoyed that Germany is still portrayed as the perpetrator country, what happens in other countries nobody talks about it, e.g. Armenia, Yugoslavia, Africa. Of course, I would never deny what happened here and I have a lot of empathy.

Her partner explained that they had never heard of Ravensbrück. He was disappointed with the exhibition: 'Images say more than words.' This couple reiterates Frevert's (2013) assertion that emotions are culturally shaped. There is a cultural expectation that one has to feel empathetic towards the victims, yet this is not necessarily how one truly feels.

An elderly group of visitors from Flensburg visited only the GDR's memorial complex, consisting of the crematorium, the mass grave and the execution path. The woman in this group was very hesitant about visiting the crematorium and was visibly distressed about the execution path. Her partner finally said:

> One attempts to step into the shoes of these prisoners, not much is here anymore but one tries to imagine what it was like. The longer you stay in these places, the more depressing they become. The contrast between the SS houses and the camp is fascinating, more should be done to integrate them. I have no idea, however, how one would present this history, obviously it should not become like Viking village where you re-live history, it is a memorial not a museum.

This visitor absorbed the sacralisation of the memorial sites in West and East Germany after the Second World War. Early commemoration at the concentration camp memorials were designed to emphasise the suffering of the victims, blending out the German responsibility for the atrocities (Knoch 2020). His partner subsequently summarised that 'original structures' such as the execution path are emotionally too much: 'one knows what to expect and tries to emotionally detach from the place'. This group of visitors shows the importance of imagination – visitors develop a new world in their heads by imagining daily life in a concentration camp which they however cannot do without pre-existing knowledge.

Alison Landsberg (2004) referred to the emotions that occur when one steps into the shoes of the victim or the perpetrator (sleepless nights in the former SS accommodation) as 'prosthetic memory'. I am reluctant to use this term. The visitor who was horrified about walking over former ashes would have not reacted in this way had she not overheard the tour guide. Moreover, considering that she complained about Germany still being viewed as a perpetrator nation, her reaction was a fleeting moment. Similarly, for 'sleepless nights' to occur the visitor must imagine himself/herself as the perpetrator, otherwise such emotions would not arise. I met one family who did not even know that they had stayed in the former SS accommodation until they visited the memorial site the following day. Thus, it is 'tourism imaginaries' that cause such emotional reactions. Patrice Keats (2005) explains that these imaginations are necessary for vicarious witnessing to occur – the visitor has to fill in the gaps with his/her own images which can derive from photographs, documentaries or books the visitors have encountered before the visit. The visit itself is then 'brought to life' by these inner images. At times, these inner images can feel overwhelming, for instance when a visitor suggests that 'the ovens are too much'. Keats (*ibid*, 181) calls such moments 'vicarious witnessing saturation point'; at such points visitors will either engage in distractions (i.e. taking photographs) or leave the site altogether.

4.1.4 Wrestling with an uncomfortable past

Reinhard Koselleck (2002) argued that Germany cannot just remember the victims. Germany has a political responsibility for the Nazi past and as such needs to acknowledge the perpetrators. When the discussion surrounding the new Holocaust memorial in Berlin emerged, there was a small window of opportunity to address the perpetrator, which was missed according to Koselleck (*ibid*). It is true that Germany only reluctantly confronted the perpetrators in the public sphere, but it is even more difficult behind the closed doors of family homes. The Israeli academic Dan Bar-On (1989) realised in the 1980s that extensive research was conducted with the children of survivors, yet there was barely any knowledge about the children of the perpetrators. He subsequently went to West Germany and interviewed 50 children whose parents were directly involved with the Nazi regime. Most of the children, now adults, had been confronted with a wall of silence, that is, they had hardly any information about their parents' involvement with the regime. Some of them described their home environment as a 'madhouse' characterised by violence. At Flossenbürg one male German visitor compared his childhood with being in a concentration camp. He was very sarcastic throughout his visit, and it later emerged that his father was involved in the SS, which subsequently led to his father committing suicide in the 1970s – an issue Dan Bar-On (*ibid*) also encountered in his research. The visitor also complained about Germany's refugee policies, arguing that this is a 'stitch-up to force Germany onto its knees'. Dan Bar-On (*ibid*) writes in his book about the frustration in hearing some of his interviewees adopting uncritically the language of the parents, so seemingly not coming to terms with the past. This visitor certainly exhibited some of those characteristics, and struggled to come to terms with a father who was clearly violent.

Moreover, Germans often became victims themselves yet could not acknowledge it due to Germany's perpetrator role, a dynamic I often encountered at the memorials sites. A distinct memory for me is an elderly couple in their 70s from the city of Augsburg/Bavaria. Shortly after their arrival in Flossenbürg, the woman explained that she had grown up in Australia where her Polish father, who had survived Mauthausen, had migrated to. She emphasised that he hardly spoke about his life prior to his immigration. She, however, married a German and subsequently moved to Germany. Her husband had a particular interest in the Jewish history of Augsburg and had written several books on the topic. He openly declared that he was a member of the Hitler Youth and was in Dresden prior to the bombings in February 1945. By chance, he was sent away from the city before the fatal night and remembers sitting on a hillside, watching the 'red inferno' destroying the entire city. He recalled this story in tears, and the long-term psychological impact of this experience was clearly visible. For this couple, the visit to Flossenbürg was a commemorative one, but also one which enabled them to deal with their own traumatic pasts. They spent an extensive period at Flossenbürg, in total six hours (interrupted by a lunch break), visiting the whole site and both exhibitions. There were also the only visitors I accompanied who watched the film *Flossenbürg* in the main exhibition hall. The woman highlighted that she always emotionally distances herself from such a visit as otherwise she would not be able to cope. At the end, she summarised her feelings: 'Although one is prepared, it is always depressing. I am always wondering how they managed to survive, I would have died on the first day.'

Similarly, a Belgian couple I accompanied throughout their visit had been coming to Flossenbürg for 14 years with the aim of commemorating the victims. The woman explained that her parents were from the *Sudetenland:* 'They never talked about their experiences, they were also not welcome in Germany.' The rest of her family had remained in what is now the Czech Republic and the woman's cousin worked at Terezin Concentration Camp memorial. They engaged extensively with the exhibition although they had been to Flossenbürg many times before. It was their way of dealing with an uncomfortable history which had affected their own lives.

Another German visitor explained to me that he had recently found a photograph in his grandfather's attic which encouraged him to conduct his own research into his grandfather's background. He concluded that his grandfather must have worked in a subcamp of a concentration camp, which explained the apprehension regarding his first Jewish girlfriend who was greeted with the comment: 'If your grandfather knew'. Gitta Sereny (2001) points out that the third and fourth generations are less angered about their forefathers' involvement in the Second World War and approach Nazi Germany with intellectual curiosity. Indeed, the current generations seek to understand Nazi Germany on a very personal level, often through genealogical research, highlighted by this visitor's experience. At Ravensbrück, I accompanied a woman whose grandfather had been imprisoned at Neuengamme Concentration Camp where he contracted a terminal illness and subsequently died at home. Her father returned home from the war disabled and never talked about his experiences. She commented during her visit that she 'learned nothing in school'. She later said: 'The atmosphere is the most important aspect of the site. When I walk around here, all these memories come flooding back, my father returning from the war disabled.' She did not visit Ravensbrück's exhibition as for her a contemplative walk across the site was more important than engaging with the historical facts of the site.

Vergangenheitsbewältigung also occurs in an international context, albeit in a different form. I encountered Belgian and American visitors at Ravensbrück who were tracing the footsteps of their grandparents. Indeed, the Belgians' grandparents had hidden an American fighter pilot, yet were reported to the local Nazi office and subsequently imprisoned in concentration camps. The Belgian and American descendants had found each other through a genealogical website and were now discovering the fate of their grandparents together. Ravensbrück was the site where the Belgian grandmother was imprisoned, and her grandson immediately recalled one of her stories: 'She was slapped in the face during one of the *Appells* and she then held her other cheek towards the officer who was taken aback.' On entering the exhibition he said: 'I can read this all at home. I just want to know where my grandmother was.' Interestingly, these visitors appeared to have been drawn to a memorial site as a way of thinking about their family's past; the site itself and its historical content played no role in these visits.

4.1.5 Negotiating competing memory narratives

Buchenwald's memorial manager, Volkhard Knigge (2004), astutely observed that all visitors might be at the memorial site, but they are not in the same place. Visitors arrive at memorial sites with ideas stemming from books, films, political discourses and/or their own family narratives (Tyndall 2004). Once the visitor is onsite, these narratives compete with each other and, whilst outwardly it might appear as if the visitor is a dark tourist by focusing on certain objects, s/he is not. Rather, s/he is seeking a validation for his/her learned narratives.

People in the GDR were used to seeing shocking images or being confronted with 'traumatic' artefacts. When they arrive at Ravensbrück now, they are bewildered. One couple from Gera/Thuringia was aware of Ravensbrück but did not know the extent of the atrocities committed. They frequently compared Ravensbrück to Buchenwald and remarked that '[w]e have seen lots of shocking images in the GDR but they are all missing today [gold teeth, piles of bodies, glasses]. Buchenwald was also more extreme in the past, today there is too much text.' Similarly, a woman from Halle/Saxony-Anhalt recalled that

> in the 70s it was awful here, really grim, one could see teeth and shoes, it's a shame you can't see these things anymore, now everything is very scientific. My son went to Buchenwald recently, he was also disappointed. I remember the lamp shades made of skin. I actually didn't want to come here again but my husband wanted to see it.

Throughout the visit of the main exhibition in the *Kommandantur*, she also often referred to books she had to read in the GDR such as *Olga Benario* or *Jakob der Lügner*.

The comments about the GDR exhibitions corresponded with the narratives I encountered during the visitor survey interviews where visitors referred to the hair at Ravensbrück, or shocking images. Indeed, the missing hair at Ravensbrück became such a frequently discussed issue that I enquired with the management team during my research what had happened to it. They explained that the hair had been on loan from Auschwitz and had been displayed in the exhibition without informing the GDR visitors of its origin. Since launching the new exhibition, the object is no longer on display due to its historical inaccuracy. These visitor comments underline the fact that, even when state narratives change, previous discourses can linger. It might even prevent people from visiting memorial sites again, as the visitor from Saxony at Flossenbürg demonstrated: 'I will never go to Ravensbrück again, just the thought of those medical experiments makes me shiver.' She was referring to the GDR exhibition and was unaware of the new museum which has reduced the use of images.

Annette Leo (2000) noted in her oral history project with former GDR citizens that the second generation of the GDR, those born into the GDR, were most likely the generation that absorbed the GDR's antifascist narrative. Unlike the first generation, who often witnessed the Second World War first-hand, the second generation had no comparison. The visitor comments confirm Matthias Heyl's (2016) suspicion that the impact of visits to memorial sites in the GDR lingers, and we therefore need to understand how this influences current and future visits to memorial sites. In 1998, Faulenbach argued that in relation to the East Germans one historical narrative should not be simply replaced by a new one, yet it is precisely what happened as there is little engagement with adult visitors at the concentration camp memorials. Thus, changing political narratives by just modifying an exhibition is not enough: there needs to be a wider engagement and, most importantly, it needs to be explained to the visitor why those exhibitions had to change.

Competing memory narratives are particularly visible amongst international visitors. As mentioned earlier, American visitors at Flossenbürg focused on the original structures and, most importantly, the memorial plaque which commemorated the American infantry division involved in the liberation. By contrast, one Swedish couple was not overly concerned with the wider context of the site, indeed one of them just enjoyed the tranquillity of Flossenbürg; she concluded that 'the site looked very beautiful' and 'every nation has done something bad'. This visitor had no difficulties with the apparent inauthenticity of the site, a very different viewpoint compared to the German or American visitors.

Increasingly, however, visits to the German concentration camp memorials are overshadowed by the 'Americanisation of the Holocaust narrative'. Jörg Skriebeleit (2016), the manager of Flossenbürg, explained that the trend of Auschwitz being regarded as the 'worst place' has been observed for a long time. Previously, it had only affected the smaller sites but more recently it has also impacted on the larger sites such as Buchenwald and Sachsenhausen.

4.1.6 Auschwitz and the Holocaust narrative as global narratives

A frequent comment during my observations was that 'Auschwitz was much worse', and/or previously visited memorials appeared to have a more profound impact. The female visitor at Ravensbrück from Emsland/North Germany, whose grandfather was imprisoned in Neuengamme, remarked that 'Theresienstadt (Terezin) was much better because you could see images in the individual cells on how the prisoners were found'. She then further added: 'Auschwitz was the most impressive one, when one sees the shoes, one can imagine that people walked in them. The hair, however, looks false.' A Polish couple at Ravensbrück argued that 'Auschwitz is more impressive as you could see the hair and the shoes'. Their Polish friend, who lived in Germany, disagreed and said: 'Auschwitz is now part of a massive marketing machinery.' Another couple from the Emsland area claimed to have had a greater emotional experience at Auschwitz: 'It is a more depressing site, one can still see the electric barbed wire fence, which confronts you straight away after entering the site.'

One teenager in a group of visitors from the Rhineland area quoted almost immediately upon arrival at Ravensbrück the film *The Boy in the Striped Pyjamas*. Hence, this teenager was already influenced by a fictional representation of the Holocaust prior to the visit. She is also typical of a new generation of visitors who no longer have a direct connection to the past and are exposed to this history through mediated presentations. How much Auschwitz, or better the associated objects, are in visitors' minds was further evidenced when one of the teenagers asked in the main exhibition: 'Whether one could also see the glass cases with the hair here?' On their walk to the former textile factory from the *Kommandantur*, another female visitor in this group asked her son 'whether Ravensbrück was equally as depressing as Yad Vashem?', to which her son responded: 'No, it was worse'; and she subsequently concluded: 'I suppose it is about the Jews.' The narrative of Auschwitz and the Holocaust is now so strong that it overshadows all subsequent visits to memorial sites. It also influences the perception of victimhood, for instance, a British couple from Kent asked me who the Sinti and Roma were, and a German visitor commented 'that it was not explained in school that there were also other prisoners than the Jews'. Reinhart Kosseleck (2002) argues that in the land of the perpetrator, we have to treat victim groups which were also murdered by the Nazi regime equally, and yet the 3.5 million Russian prisoners of war, of whom 60 per cent were starved to death (or shot as in Flossenbürg), do not have a memorial. This research supports Koselleck's concerns about a general lack of awareness of other victim groups.

The memorial site of Auschwitz-Birkenau was also perceived to be worse at Flossenbürg. Four young people I accompanied displayed very little interest in the overall site. Indeed, when we arrived at the crematorium, one of the visitors asked me: 'What did they do in there?' On leaving the site, one of the young women commented that

> Auschwitz [which she had visited on a school trip] was much better as you could see the scratch marks on the wall, the mountains of hair, glasses and baby shoes which is more exciting. You can even see a mass grave at Auschwitz.

Although there is a mass grave at Flossenbürg, it fades into insignificance compared to Auschwitz. These comments confirm Aleida Assmann and Juliane Brauer's (2011) argument about the sense of voyeurism young people develop when constantly confronted with violence in a mediated world.

Christian Kuchler's (2021) research into German school trips to Auschwitz in the 1980s revealed that pupils often spoke of shock when encountering the hair, the glasses, suitcases and/or prosthetic limbs. He concludes that the educational potential was limited as the confrontation with the artefacts overshadowed the entire visit. I argue that we already have entered a phase where we remember the artefact rather than the historical event. For instance, a group of visitors who had booked a small guided tour at Flossenbürg commented after the visit that 'I could not eat anything after visiting Auschwitz, but here I quite happily go to the café'. What these examples reveal is Oren Stier's (2015) notion of remnants. The hair, the shoes and the glass cases stand for the human suffering, and once these objects are removed, the memory of the event itself fades.

A family comprising three generations who visited Flossenbürg demonstrated poignantly how the generational change affects Holocaust memory. These visitors consisted of grandparents (German), daughter (German) with her husband (American) and two teenagers (American), one of them the grandson. The exhibition did not capture the teenagers' interest, whereas the grandparents were actively engaging with the content. The grandfather revealed that he had lost his father in Flossenbürg. Indeed, he commented in a rather matter of fact way that his father 'died on 1 November 1943, probably either shot or due to malnutrition'. As they watched the film of the liberation including the mass exhumation, the grandfather moved away from the video in disgust. When the family later arrived at the crematorium, the grandmother initially refused to enter it, while the rest of the family embarked on taking photographs. The grandmother also struggled to comprehend the ramp which connected the camp with the crematorium. By contrast, the teenagers were less concerned with the potential atrocities that had occurred onsite, including the great grandfather's perishing at the camp. They asked me whether the crematorium was a gas chamber, and when I answered 'No', they proceeded to question me about the number of gas chambers in Auschwitz and the difference between cremation and gassing.

For some visitors, the wider historical context of the Nazi past has been reduced to the gas chambers and barbed wire fences in Auschwitz. My research has demonstrated what various scholars (Stier 2015; Crane 2008; Ebbrecht 2010; Cole 1999) have already highlighted: the domination of Holocaust icons in popular culture leads to fetishisation of objects such as crematoria, gas chambers or hair in glass cases. In particular, Flossenbürg, the site with the largest percentage of American visitors, has revealed this fetishisation. The most popular locations for photographs were the crematorium, the mass grave (ash pyramid), barbed wire fence posts, watchtowers and the American memorial plaque, which leads to what Barbara Zelizer (1998) calls 'remembering to forget'. The focus on these objects prevents visitors from understanding the wider context. For instance, at Flossenbürg, the quarry was the location where death often occurred as prisoners worked under the most horrific circumstances, not at the barbed wire fence posts or at the watchtowers.

The visitor statements are also perfect examples of what Norman Finkelstein (2014) calls 'unique suffering'. No other pain can be compared to the Jewish pain. Creating this hierarchy of suffering is one of the greatest challenges of current commemorative practices of Nazi history. Indeed, Volkhard Knigge (2010) criticises the German memory culture that is now dominated by hollow ritual events, and the belief system that remembering the Nazi past will automatically strengthen democracy in the present. These examples highlight that a memorial visit will not necessarily shift pre-learned perceptions. The globalised Holocaust memory culture has, in fact, eroded a more nuanced understanding of the Nazi past.

4.1.7 The gendered 'gaze'

Ravensbrück was the first memorial site that 'installed a path-breaking exhibition' (Jacobs 2010) on the sexual exploitation of women in concentration camps. For the memorial team, this required walking a very fine line between respecting the victim, not objectifying women yet again, and not overburdening the visitor. Interestingly, it was often women who engaged with this part of the exhibition intensively. For instance, two teenagers were very quickly exasperated by Ravensbrück's extensive exhibition, but when they reached the room of 'sex slave labour' their attention changed and they watched the eyewitness accounts. Their mother later commented on the women's experiences: 'I cannot believe how cold hearted they were, conducting forced sterilisations without anaesthetics.' Similarly, two young women from Germany and the United States spent most of their visit in the room that displayed the prostitution in concentration camps, reading files with personal accounts. By contrast, one German male visitor said that he was distressed by the sexual violence: 'sexual assaults are a particularly difficult aspect to display in the museum', while another two German male visitors skipped this part of the exhibition altogether, commenting that this is 'too much'.

While some visitors were challenged by the female experience at Ravensbrück, others appeared to be under the impression that the female experience of Nazi imprisonment was less gruesome. The Polish visitor who criticised the dominant discourse of Auschwitz commented that 'he does not understand why there was a women's concentration camp, since they are the nice sex and just have the babies', while the British couple suggested, after seeing the striped dress, that 'this does not look too bad' and 'we hope that the conditions better here than in Auschwitz'. Even the American and Belgian visitors who traced the footsteps of the Belgian grandmother thought that 'the women were treated better' and 'Buchenwald will be much worse'.

Zoë Waxman (2017) argued that a feminist perspective of Holocaust experience is absent, since women are absorbed into generic victim groups such as the Jewish or the Jehova's witnesses, even though female experiences were distinctly different. For instance, no longer having your menstruation caused worries about future possibilities of motherhood, and women who still had their monthly cycle faced the degrading experience of no sanitary products. In fact, one female visitor at Ravensbrück commented that she did not know that malnourishment can cause amenorrhea. Women with children and pregnant women were also often immediately destined for death, thus making them more vulnerable than men who could be exploited longer for forced labour. One of Ravensbrück's largest victim groups were the so-called 'asocials' whose crime was a sexual relationship out of wedlock and/or any other supposedly inappropriate female behaviour. The women experienced an additional stigma within the camp since they might have caused the imprisonment with their 'promiscuous' behaviour. Only in 2019 were the 'asocials' recognised as an official victim group by the German government, too late for many. In addition, women were more prone to sexual violence and/or forced prostitution in camp brothels, making them double victims. Yet it is a history that is neglected, and the visitors' comments are therefore not surprising. Nicole Bogue (2016) points out that Auschwitz I and Auschwitz-Monowitz do not indicate the existence of camp brothels in the museum and/or at the memorial grounds. Thus, for visitors at Ravensbrück the sudden confrontation with the *Sonderbauten* might come as a shock. These visitor reactions certainly highlight the need for further research into exhibition practices of the female experience of Nazi imprisonment and of sexual violence and their impact on visitors.

4.2 Responses to the perpetrator – the House of the Wannsee Conference

Germany only reluctantly dealt with the perpetrators of the Second World War, and in both East and West Germany, the pictures of only a few selected 'beasts' initially emerged (Pearce 2011). It was not until the publication of Goldhagen's *Hitler's Willing Executioners* in 1996 and the *Wehrmachtsausstellung* in 1999 that the perception of an exceptional few perpetrators began to change. At the concentration camp memorials, perpetrators were often not exhibited. Ravensbrück launched its first exhibition about the perpetrators in 2004, followed by revised exhibitions in 2010 and 2020. Neuengamme Concentration Camp memorial followed in 2005 and Sachsenhausen Concentration Camp memorial in 2017. At Flossenbürg and Ravensbrück, visitors often asked three fundamental questions: How could this continue for six years without intervention? How did the local community react to the concentration camp? And why was the prosecution rate of the perpetrators so low? While memorials highlight the links between the camp and the local community, the other two questions often remained unanswered. Günter Morsch (2018), Brandenburg's head of memorials until 2018, noticed this shift in visitor behaviour and argued that exhibitions need to take a closer focus on the perpetrator.

The House of the Wannsee Conference filled this lack of focus on the perpetrator initially, yet worried that the site could become a place of pilgrimage for the far-right. Thus, during my research the perpetrators were presented alongside personal destinies and a comprehensive history of the Third Reich. Violent images, as shown in the first exhibition, were reduced to avoid an overburdening of visitors and a secondary victimisation. Nevertheless, as the visitor research has shown, emotions were very strong at times. In fact, the results of the visitor research at the House of the Wannsee Conference surprised me. The site itself is located at the beautiful Wannsee Lake, which was often even more stunning during my research at the height of the summer in August 2016. The villa contains no traces of violence, so it was an ideal location to observe visitors at the least authentic site. However, visitor research was often made difficult due to the sheer volume of visitors at the site, making it challenging to observe visitors closely and/or to conduct in-depth interviews. Moreover, unlike my other research sites, the House of the Wannsee Conference had the largest international audience, yet at times I had to exclude visitors whose English or German language skills were not sufficient to conduct interviews.

Although German visitors were still the largest visitor group, this was followed by visitors from Israel, the United Kingdom, the United States and the Netherlands. The majority of visitors surveyed were in the age group between 61 and 70 followed by the age groups 51–60 and 41–50. The sample size included a higher percentage of male visitors compared to Flossenbürg and Ravensbrück, and a bachelor's degree was most frequently stated as the highest qualification. The visitor interviews also revealed a greater religious diversity compared to my other two studies, Ravensbrück and Flossenbürg, with Catholicism and Protestantism the most reported religious affiliations followed by Judaism. Overwhelmingly visitors responded that they did not have a personal connection to the site; however, this question caused upset amongst visitors from Israel. They felt that this question was unnecessary as their personal connection was unquestionable. On further examination, it transpired that the grandparents' generation was often a victim of the Holocaust, hence current Israeli visitors identified themselves as 'secondary victims'. Subsequently, for these visitors the main motivation for visiting the memorial site was a sense of duty. Most other visitors stated that they wanted to acquire more knowledge or just had a day out in the Wannsee area of Berlin. Although visitors overwhelmingly reported that they had prior knowledge of the House of the Wannsee Conference, further questioning revealed that this was limited to the concept of the conference or the final solution, which can be explained by the stronger media influence. Unlike Flossenbürg and Ravensbrück, visitors referred more frequently to films (e.g. *Conspiracy*) and documentaries as an influential factor in visiting the site. In this regard, Wannsee differs significantly from the concentration camp memorials where visitors barely referred to popular culture as a motivator for the visit.

Wannsee's participants in the visitor observations reflected the international audience with most of them still from Germany (14) but also from Britain (3), Austria (1), The Netherlands (2), Belgium (1), the United States (1), Australia (1) and Israel (1). The age groups 50 to 60 and 60 to 70 were the dominant ones during the visitor observations, yet visitors in the age group 30 to 40 were also part of the sample. In general, the ethnographic research at Wannsee was difficult since most visitors did not spend much time onsite, leaving a smaller opportunity to create a trusting relationship with the visitors. Moreover, overcrowding in the rooms interfered at times with the close observations of the visitors. Nevertheless, like at Flossenbürg and Ravensbrück, the observations revealed the unique visitor responses to the site.

4.2.1 Impact of the exhibitions and the overall memorial site

The questions which were designed to capture visitors' engagement with the site highlighted that the vast content of the exhibition overwhelmed visitors. Although visitors responded positively to the question 'Do you remember objects/images which represent suffering', no common themes emerged; they mentioned a variety of images and objects ranging from 'mass execution' and 'race ideology' to Jews digging their own graves. The question regarding the most memorable part of the exhibition produced similar results, with the most common words used being 'mass execution', 'protocol' and 'conference room', which are generic and could refer to several parts of the exhibition. Nevertheless, 'mass executions' and 'Jews digging their own graves' refer to a section of the exhibition that is particularly graphic. Like Flossenbürg, it is these violent images that stay with visitors.

Visitors agreed that the visit provided them with more information, inspired them to visit similar sites, and encouraged them to engage with the Nazi past afterwards. Nevertheless, like Flossenbürg and Ravensbrück, a significant proportion reported that they had already engaged with the history, irrespective of the visit. Interestingly, this sentiment was not only shared amongst German visitors but also amongst French, Dutch, Israeli and US visitors (35 per cent). Bearing in mind the context of the exhibition and the background scenario of my fieldwork (shortly after the Brexit vote, ongoing refugee crisis in Germany, forthcoming elections in the United States), the visit still did not seem to question visitors' own behaviour as 54 per cent disagreed with the statement.

The most surprising answer was in response to the question 'Do you think, the site represents a special atmosphere', with the vast majority of visitors saying 'yes'. The reasons for this response were 'the oppressive atmosphere', 'the conference', 'the beautiful house versus the terrible decision', 'depressing' and 'chilling'. This stands in stark contrast to the results of the two concentration camp memorials where visitors equated the lack of historical structures with the lack of a sense of 'aura'. As mentioned previously, I conducted my research at the height of the summer in August 2016, which was often sunny and very hot. Wannsee itself is also the least authentic site, and yet visitors described it as chilling and oppressive. In essence, visitors projected their emotions about the perpetrator onto the building. It is the 'indexical authenticity' – the link between the building and its past – that creates these feelings of authenticity. In addition, Gerd Kühling and Hans-Christian Jasch (2017) explain that with the intensification of perpetrator research in the 1980s in West Germany, emotionally charged projections onto the Wannsee Conference increased. The press, for instance, frequently describes the villa as 'an idyllic place of horror'. Moreover, there is a widespread public perception that the murder of the European Jews was decided here, heightening the 'evil nature' of the house. Visitors' description of the visit reflect the emotionally charged perception of the Wannsee Conference, with the most commonly used words to describe their inner state of mind being 'depressive', followed by 'impressive, informative, moving, interesting and emotional'.

4.2.2 In search for authenticity and an emotional experience

Wannsee was the site of stark contrasts. On the one hand, I was confronted with visitors who were predominantly interested in the historical authenticity of the site, often derived from films, books or previous visits to other memorial sites, and on the other hand I encountered visitors who responded very emotionally. A couple from Nottingham initially spent time taking photographs of the 'Mein Kampf' poster, and of women with banners in their hand saying 'I am a racial dishonour'. The man subsequently commented that he recognised a lot of those images. In the conference room he took photographs of the protocol and suggested that 'the table should be here, the documents are amazing'. He also commented how unbelievably matter of fact the protocol was while recalling the names of Stuckart and Kitzinger in the film *Conspiracy*. At the end of the visit the man concluded:

> For me less is more, authenticity is the most important thing and this place has it. I can't believe I'm actually here. I know all about it, I have read so much about the Nazi period. I've been to Auschwitz and the Anne Frank House. Auschwitz I have to go again to see it in the winter, I went in the summer. The Anne Frank House was chilling apart from the shop at the end. In room 11, things started to get too much for me, so I was only picking the things I was interested in.

Similarly, a single male visitor from Belgium was clearly disappointed about the lack of authenticity, commenting that '[t]hey need to sort out the conditions in this house, so that they can show the original document' [due to climate control issues the original document could not be displayed]. He took photographs of the Nuremberg racial laws and the Evian conference caricature, but regarded the rest of the exhibition as boring. He was extremely fascinated by the Second World War and has seen many films and memorials; indeed, he had already spent four hours at Sachsenhausen.

A family from Austria explained to me after the visit that the interpretation of the Second World War in Austria had improved, and the daughter had already visited Mauthausen. The male visitor in the group remarked that 'it is important to be in an authentic space', which, however, was contradictory by his behaviour as he barely spent any time in the conference room. By contrast, a single male visitor from New York walked straight into the conference room and asked me 'Is this the Dining Room and are these the same floors?' He was disturbed by a large school group and therefore involuntarily walked through the other rooms, but was evidently not interested. He returned quickly to the conference room, read all the documents and took photographs of the room and the lake. At the end of the visit he told me that he had a whole library of books about the Second World War at home and recognised all the images. He had already been to the Holocaust memorial, the Topography of Terror, Auschwitz and Sachsenhausen. And he had watched 'the two superb films *Wannsee* and *Conspiracy*', which motivated him to visit the memorial. Derek Dalton (2019) admits that he immediately had the images of the *Conspiracy* film in his mind as he approached the villa during his stay in Berlin. He also wondered whether other visitors would make similar associations, which this research certainly suggests. Nevertheless, one can observe the references to fictional representations of the Holocaust mainly with visitors from outside Germany as during my observation I did not encounter any German visitors who referred to the films.

In line with the quest for authenticity is the need for an emotional experience. For instance, a woman from the United Kingdom missed the final parts of the site and sat outside with tears in her eyes. She commented: 'The personal stories grabbed me most, not the facts. The people you meet at the beginning, it's a shame it doesn't follow through though. You have to have an emotional experience at these places, otherwise it makes no sense.' Her husband, however, concluded that the first time he visited the House of the Wannsee Conference 'he had goose bumps, but now that feeling is gone'. A couple from Berlin was particularly affected by the Eichmann trial and the woman commented 'unbelievable'. Although her partner suggested that he expected more about the conference itself, he says that 'when one engages with this exhibition, then one has one shiver after another one running down the spine'. An Australian visitor mentioned that she had read the novel *The Book Thief* which made her want to understand how a nation could descend into such disaster, and she therefore photographed the majority of the election posters and the description of the Nuremberg racial laws. She spent a long time in the various exhibition rooms, but when she encountered the section about the extermination camp, she started to cry. Thus, whilst visitors might desire an emotional response, it can easily turn to emotional fatigue. Hence, memorial managers have to walk a very fine line between engagement and exhaustion, demonstrated by the following couples.

Two couples from Berlin briefly looked at the conference room before leaving the site again. Afterwards one of the women informed me that she had been a lecturer in cultural studies and she felt that the exhibitions in the GDR were more interesting (the lamp shades in Buchenwald, the ash in the lake Ravensbrück), but today 'the memorials have been sanitised as one cannot confront visitors with such extreme images'. One of the men in the group commented that 'the presentation of the house's history is poor', yet he still believes that 'it is important to read the documents (e.g. deportation documents) in an authentic space'. Another woman in this group summarised the visit as follows: 'it is a nice house in such a gruesome space'. These couples highlighted again that the GDR's shock pedagogy has long-lasting after effects which influences visitors to this day and might even have caused 'numbness', that is, today's less emotional exhibitions are experienced as dreary. On the other hand, these couples confirm Kühling and Jasch's (2017) assertion that the Wannsee Conference is emotionally charged, which impacts on people's perception of the villa.

4.2.3 Individual Vergangenheitsbewältigung

Whilst most international visitors were keen to encounter the 'authentic traces' of the Wannsee Conference (and the evil perpetrator), for German visitors the site could lead to intense questions about one's own family involvement with the Nazi regime and Germany's past as a whole. I accompanied two women in their 30s who left the exhibition twice in tears. They were both politically active in the trade unions in Germany. They related very deeply to the content of the exhibition and commented that the racism is repeating itself in current society. In room three, which discussed Jewish life in Germany in the 1920s, one of the women commented: 'For me, *Vergangenheitsbewältigung* only commenced 15 years ago. My grandmother had the Swastikas everywhere in the house, she was a lovely woman and yet she was convinced of the Nazis. My grandfather was also a member of the NSDAP.' In room five, one of the women could not cope with the images of the mass shootings and said that she had to leave, saying: 'I feel sick.' In room eight, which explained the invitations to the Wannsee conference, one of the women remarked: 'This room is worse than the pictures on the wall.' When they subsequently read the Eichmann trial file one of the women started to cry. At the end of the visit, one of the women explained that one set of grandparents was involved in the SS, but her father never talked about it, whilst the other set of grandparents were hiding Jews in the basement. Her history stands for so many grandchildren generations who are trying to come to terms with the void left behind by previous generations. She, however, is also representative of the inner conflict of subsequent generations who remember their grandparents as loving individuals who at the same time supported a criminal regime.

Another example was one female German visitor who commented at the end of the visit that it is

> somewhat different to be in an authentic site, I have cold shivers running down my spine. My mother was a secretary in the *Wehrmacht* and she claims to this day that she did not know anything. And she still complains about the Jews and the immigrants, it makes me furious. I will never understand what is/was going on in their brains.

Her partner said that it was an interesting exhibition, but there was too much content, and he probably will have forgotten everything by the time he leaves the site, suggesting that he 'can read everything on the Internet today'. He said that he is 'amazed that nobody resisted' to which his partner replied that 'they were all afraid'.

A single female visitor from Berlin arrived at the memorial site very early, explaining that she wanted to avoid the crowds, and she certainly did not want to encounter any school groups. She spent an extensive time at the memorial and told me that she was a teacher and had been to many concentration camp memorials: 'It is the worst chapter of German history.' At the end, she asked to be left on her own in the garden, saying that 'one has to be emotional when you leave the site here, it could not be any different'. For her, this visit was a deeply personal one which involved an active confrontation with Germany's responsibility for the Nazi regime.

By contrast, a German male visitor explained that he worked for the *Bundeswehr* and for him the role of the *Wehrmacht* was particularly interesting. He often discussed it with his grandfather and was very conscious of the tension between Germany's past and the current *Bundeswehr*. He knew a lot about the content of the exhibition as he had already been to Buchenwald and Bergen-Belsen; consequently the exhibition did not affect him emotionally. He, however, emphasised that it was great to see an exhibition of the whole Nazi history from the start to the end.

Visitors from Israel formed one of the largest visitor segments at the House of the Wannsee Conference. However, language often caused a barrier to a meaningful engagement with them, except for one couple: he was from Israel while his wife was German. He explained to me that his grandfather was from Lithuania and had fought for the Russians during the Second World War, did not return to the Soviet Union and went to Israel instead. His grandmother was from Leipzig and managed to flee; she survived and finally went to Israel. The rest of the family had died in the Holocaust. They both spent an extensive time in the first part of the exhibition, discussing the racial laws. The man was shocked when he encountered other German visitors who were laughing in the exhibition. He subsequently listened to Hitler's speech on 30 January 1933 and read the personal letters from Eichmann. When he encountered a transportation order for Jews from Stettin (today Szczecin), he asked his wife 'What would you have taken?' He also spent a long time listening to the Eichmann trial and read the corresponding files. At the end, his wife said that it was an interesting exhibition, while he explained 'that it is too much text. It is important to have an emotional experience at these sites which you will not have with copious amounts of text.' He then compared the House of the Wannsee Conference with Yad Vashem where he was a tour guide. This visitor had a very personal connection to the site, and the rather clinical exhibition design and the behaviour of German school-children onsite for him felt offensive, similar to Yair's (2014) analysis of German Holocaust exhibitions.

4.2.4 Repetitive exhibitions/images

Wannsee was also a site of stark contrasts in relation to visitors' judgments of the exhibitions. A recurrent theme was the repetitive nature of the exhibition, and that images had already been seen elsewhere. For instance, a couple in their 50s from the United Kingdom commented that they had seen the images in other exhibitions, and similarly a visitor from Holland, who was a history teacher, remarked that the images on display are generally known. Two German history teachers commented that the majority of images can be found in every text book; the most famous one of the young boy with his mother in the Warsaw Ghetto. In line with these sentiments was a general feeling of 'saturation', since visitors remarked that they did 'not come here to learn about concentration camps'. This particular comment was from the Jewish visitor from Israel; however, it was shared by numerous other visitors, for example, a male German visitor in his 70s said: 'The exhibition is depressing, I have already been to Dachau and Mauthausen and therefore I do not want to see anymore', albeit emphasising that he 'learned nothing in school, it was all supressed'. Equally, two women from Germany felt that the exhibition was 'too much'. They had already been to Neuengamme, Stutthof and Plötzensee and remarked that it is often repetitive, with one of the women insisting that she 'doesn't need to see concentration camps anymore, that's enough now'.

A German history teacher seemed to get increasingly angry throughout his visit: 'As a history teacher I can't learn anything here. It's very abstract here [the conference room]'; and he later commented: 'The exhibition screams of the 1970s; it's guilt driven and that's why I have to exhibit everything. The exhibition could have been anywhere; it does not make a difference.' Visitors also had the expectation of engaging with the conference itself rather than with the whole history of the Third Reich. So commented a single woman in room nine [conference room] that 'this is the room I came here for', but 'this exhibition is not for me; I'm more into personal stories, they affect me. A lot of the things in here I already knew. I have already reached saturation point from my school days'. Similarly, a family from the United Kingdom explained that they had already been to Sachsenhausen (which was very emotional) and Hohenschönhausen (which was brilliant) and that they were therefore 'only' interested in the conference, skipping the final parts (extermination camps) of the exhibition.

What these examples highlight is the difficult position of the House of the Wannsee Conference within the wider memorial landscape in Berlin, and indeed in Germany. By the time visitors explore this memorial, they have already been to several sites with similar content, and they therefore do not want another explanation of the Nazi period. Furthermore, the layout of the exhibition led the visitor throughout the building in a figure of eight, with the conference room being the final part of the exhibition. Yet by this stage visitors were overwhelmed by the vast content and were no longer prepared to engage with the conference itself. One family from Holland summarised it as follows:

The exhibition is okay, I was interested in the Wannsee Conference itself, so I have spent too much time at the beginning [and consequently he did no longer have time for the conference part of the exhibition]. One can see here that they were normal people. At the end of the day, it is just important to have been at this place.

Hence, a visit to a memorial site is not necessarily about engaging with history, the value lies in 'being there'. Nevertheless, while some visitors seemed exasperated by the vast content, for others exhibits about the rise of Nazism and the race ideology sparked debates about contemporary politics.

4.2.5 The tourist as a critical thinker?

The first part of the exhibition detailed the rise of the Nazi ideology in Germany and Europe in the 1920s and 1930s, the Weimar Republic and the beginning of the Jewish prosecution in Germany. This part of the exhibition sparked visitors' interest and often resulted in debates between them. For instance, a female visitor from the United Kingdom pointed out Nazi propaganda 'We call for the state to take care of its citizens first', and remarked: 'Interesting, in an economic crisis the NSDAP had more votes; you only have to exchange the word Jew for the word Muslim.' One elderly German man remembered how his father grew up with the narrative that the 'Jewish had murdered Jesus' and how the Christian churches had encouraged the hatred against Jews. One Dutch male visitor was particularly shocked about the interpretation panel on racial laws. He discussed with his friend the Nuremberg laws, and in front of the November pogrom he commented: 'And nobody has questioned it.' Hitler's quotes about anti-Semitism and the First World War sparked further debates and he remarked: 'It is depressing what they came up with, and we still see it today, for instance in Yugoslavia.'

A couple from Berlin in their 50s spent a long time in room three. They were in disbelief about the election posters and the advertisement for the book *Mein Kampf.* They further remarked that the poster 'Deutschland den Deutschen' [Germany for the German] is still visible in Germany today'. Subsequently, the man commented that 'I got sick and tired of the Nazi period during my school years, but seeing it here again, it does blow me over; it is incredible what they came up with'.

In light of the refugee crisis in Germany in 2015 and 2016, the memorial site decided to change one of the interpretation panels by including a section about the Evian conference in 1938, which resulted in a refusal to accommodate Jewish refugees by the majority of the participating countries. The caricature used was from a newspaper article in the *New York Times* in 1938 which sparked discussions amongst some visitors. For instance, two Dutch men discussed the issue of Jewish emigration, and a German visitor with his niece spent some time debating the situation of refugees in contemporary Germany.

A conversation with the Head of Education at the House of the Wannsee Conference revealed that the memorial did not encourage comparison with present day political situations such as Erdoğan in Turkey or the election of Trump in the United States. According to Elke Gryglewski (2016), a parallelisation of historical events (e.g. comparing 1920s Germany with the present day) would be a dangerous move as one cannot compare these situations; in her view Hitler gaining power in Germany was the result of a unique set of circumstances which cannot be transferred to the present. Thus, in 2016 the House of the Wannsee Conference discouraged comparisons with present-day political situations, and yet visitors used the exhibition to discuss contemporary politics. Hence, a memorial site cannot influence how visitors respond to an exhibition, nor should they pursue such an aim. Freeman Tilden (1977, 9), the founding father of the modern concept of heritage interpretation, argues that 'the chief aim of interpretation is not instruction, but provocation', and 'information, as such, is not interpretation'. He further (*ibid,* 9) adds that 'any interpretation that does not somehow relate what is being displayed or being described to something within the personality or experience of the visitor will be sterile'. These examples have highlighted that visitors want to be stimulated and not, as frequently stated, be confronted with images and textual information they have encountered at numerous other sites.

4.3 Tourism and transitional justice – visitor experiences at the former Stasi prison Bautzen II

Memorial sites are regarded as an essential part of the transitional justice 'toolkit'. While truth commissions, tribunals or reparations intend to provide justice for past wrongdoings, memorials are thought to 'heal' the nation through ongoing dialogues and debates (Buckley-Zistel and Schäfer 2014). Moreover, most memorial managers hope to create spaces for reflections, mutual understanding and respect. In essence, the memorial sites' task is to acknowledge the past but also look into the future to ensure a 'never again'. In Germany, however, the legacy of the GDR is a contested one, with former GDR citizens being accused of nostalgia or amnesia if they highlight positive aspects of their life. It has led to a memory culture that focuses on the 'brutal' aspects of the GDR regime and neglects the 'softer' parts of everyday culture. The memorial sites play a significant role in sustaining the narrative of the *Unrechtsstaat* (unlawful state), in particular Hohenschönhausen with its focus on eyewitnesses as tour guides. Visitors are encouraged to sit in interrogation chairs, or are suddenly shocked by the slamming of doors, with the aim of reinforcing the GDR as an unjust state (Buckley-Zistel 2014). This form of memory culture has come under sharp critique after the dismissal of the former director Dr Hubertus Knabe. Habbo Knoch (2020) explains that whilst there is a consensus at the memorial sites of the Nazi past that an overburdening of visitors is inappropriate, at the memorial sites that commemorate the GDR this consensus is absent. It is common to re-stage prison interrogations in an attempt to emotionally draw visitors.

Whilst academics have long focused on Hohenschönhausen and its shortcomings, the visitors themselves are neglected. For instance, Sara Jones (2014) refers in her book to the visitor comments made in Hohenschönhausen's annual statement. However, these need to be considered carefully as they were selected by the management team to justify meeting its targets. In general, memorials tend to be analysed according to the messages they convey and/ or their overarching management; the social meaning for the visitors of the sites is woefully neglected. In fact, Light et al. (2020) point out that it is the tourist who will engage with the sites, and yet transitional justice studies have not considered tourism activities and *vice versa*. Their research at a former Communist prison in Romania revealed the complex reactions of domestic tourists. The visit could result in either confirming what one would have already known or in emotional exhaustion. Thus, whether memorial sites fulfil their original intention of providing justice while also creating space for healing remains questionable. Indeed, in the case of the Stasi prison Bautzen II, I would argue that it can even entrench the division between the former East and West Germany.

Like in Romania, most tourists at Bautzen II were from the surrounding area, so it is predominantly a location of domestic tourism, with 35 per cent being from the federal state of Saxony followed by the other Eastern states of Thuringia, Brandenburg and Saxony-Anhalt. Visitors fell largely into the age groups of 51 to 60 and 41 to 50. For 80 per cent of visitors, it was the first visit, and 93 per cent claimed that they had prior knowledge about the site, often stating that it was a Stasi prison. In contrast to the other memorial sites, a higher percentage of visitors at Bautzen had a personal connection (13 per cent). Nevertheless, this was not often necessarily a family member. Visitors mentioned that they had a friend or knew a neighbour who had been imprisoned at Bautzen II.

The main motivations for visiting the site were a desire for learning (16.9 per cent), curiosity (10 per cent), wanting to see the true place (11 per cent) and finding out about what really happened (48 per cent). Indeed, a lot of visitors from the former GDR emphasised that they were born in the GDR and therefore it is *their* history. Other influential factors were the media with 22 per cent stating that they had watched a documentary and 22 per cent mentioning that friends or relatives motivated them to visit Bautzen II. Unlike at the concentration camp memorials where visits occurred spontaneously, at Bautzen II visits appeared to be planned. Indeed, some visitors stated that they had wrestled with the idea of visiting the site for some time. The GDR is clearly still a strong part of people's lives, and a visit to Bautzen is an attempt to come to terms with its legacy, but also to finally understand what happened behind the closed doors of the GDR's most notorious Stasi prison. A common phrase in the GDR was 'be quiet or you'll end up in Bautzen', as Bautzen was generally known and very much feared.

Since the former East Germany is largely a secular society, most visitors (59 per cent) stated that they did not belong to a religious affiliation. The bachelor's degree (35 per cent) or *Realschulabschluss* (35 per cent) were mentioned as the highest qualifications. At Bautzen, I did not encounter any rejections for the ethnographic research, thus highlighting that many visitors were keen to talk about their experiences of visiting the site. Since Bautzen is mostly a site for domestic tourists, all visitors who took part in the observations were from Germany and mainly in the age group 50 and above.

In contrast to Hohenschönhausen, Bautzen allows its visitors to explore the site by themselves, but offers guided tours for visitors on weekends or pre-booked for group visitors. These guided tours are conducted by trained staff and not eyewitnesses, except for the annual heritage open day where former prisoners are onsite to answer questions. One open day occurred during my fieldwork, which therefore influenced the opinions on the guided tours. Of those visitors surveyed, 13 per cent took part in a guided tour, which were very positively received. Visitors were impressed with the vast knowledge of the tour guides, the personal stories and the ability to talk to former prisoners (during the open day only). But even without eyewitnesses, Bautzen's strong focus on personal destinies in the exhibition mostly does not fail to make an impact.

4.3.1 Impact of the exhibition and the memorial site

Visitors particularly remember the personal destinies but also the overall condition of the site. Most distressing for visitors were the solitary confinement cells, personal biographies and the 'tiger cages'. 'Tiger cages' were cells within a cell: the sanitary facilities were separate from the rest of the cell, which meant that the prisoner could see the facilities, but could not use it. It was a form of psychological terror, used to break down the prisoner's spirit.

Although over 60 per cent of visitors stated that they recall objects or images of death and suffering, on further questioning they often could not name these objects and said 'it is the whole site'. The question whether the memorial represented the 'true atmosphere' resulted in a surprising answer: 54 per cent agreed while 45 per cent disagreed (one did not know). Although 45 per cent disagreed, this was not because they felt that the site was historically 'inauthentic'. They stated frequently that it would be impossible to display the psychological terror: the sounds were missing (dogs barking, keys) and the prisoners themselves were no longer present. Hence, there was a strong awareness that the conditions of the prison must have had an enormous impact on the inmate. In addition, visitors identified themselves with the victim in that they showed empathy with the emotional state of the prisoner which could not be exhibited. Visitors also frequently commented on the narrowness of the prison cells, which seemed to evoke feelings of claustrophobia. Of those visitors who had agreed that it was the true atmosphere, they stated that they can imagine what it must have been like or 'everything was left as it was'. The idea that 'not much has changed' is fascinating as some of the prison cells have been reconstructed to highlight the change in conditions between the 1950s and 1980s, which is explained to visitors. Furthermore, Bautzen also documents the history of the Soviet Special Camp located at Bautzen I, and has therefore reconstructed a cell of the former Soviet Special Camp to demonstrate the inhumane conditions there. Some visitors, however, commented that Bautzen felt 'inauthentic' in comparison to Hohenschönhausen or other prisons such as the sites in the Baltic States. In fact, the remark about the prisons in the Baltic States was made by Dutch visitors for whom communist crimes do not possess a personal meaning, and thus are seen from a position of a detachment.

Visitors' opinion was divided on the statement whether the visit encouraged them to find out more about this aspect of German history as 49 per cent either agreed or strongly disagreed, while 41 per cent either disagreed or suggested that they had always done it. Similarly, visitors either agreed or strongly agreed with the statement that the visit encouraged them to engage with similar sites, while a significant proportion disagreed or commented that they had always done so. Indeed, one visitor claimed that he did not want to see these sites anymore. The expectations many visitors had about the site were fulfilled as visitors overwhelmingly agreed that the memorial provided them with a greater insight into the Stasi prison system. They, however, objected to the notion that the visit made them question their own behaviour with 59 per cent disagreeing, 22 per cent agreeing and 11 per cent being unsure. A comparison between the federal state and this opinion statement revealed that this opinion is irrespective of where visitors come from; visitors from the former GDR shared this sentiment with visitors from the former West Germany. Indeed, visitors from the former GDR often emphasised that they 'had never anything to do with the Stasi'. This is an unsurprising statement since Germany's memory culture is focused predominantly on the Stasi, and having worked for them in either an official or unofficial capacity is a stigma in someone's biography. The statement whether the visit had allowed them to work through their own experiences also divided opinion with 43 per cent agreeing and 46 per cent disagreeing. The comparison between the visitors from the former GDR and the former West Germany also highlighted no significant difference. In fact, the cross tabulation showed that even visitors from Saxony, Bautzen largest visitor group, were divided in responding to this question; while some seem to relate to the experience of the Stasi prisons, others reject it.

Although the visitor survey research has revealed a stronger engagement with GDR films and documentaries, the highest proportion (30 per cent) did not watch films or documentaries on a regular basis. They also did not visit exhibitions or read books about the GDR, so one could conclude that there is no particular interest in GDR history beyond the visit to the memorial site. Although visitors were planning to visit other memorial sites, they often could not name specific ones. The visitors who had definite plans for future destinations mentioned Hohenschönhausen, Hoheneck, Chemnnitz-Kaßberg or the Grenzmuseum DDR (border museum GDR). Hohenschönhausen and the border museum (of which there are various across Germany) would be the only sites currently accessible. Hoheneck, the GDR women's prison, and Chemnitz-Kaßberg are currently in development.

4.3.2 Emotions

Visitors described a whole range of emotions after the visit, ranging from shock, compassion, anger, sadness and gratitude to apprehension. Hence, visitors' inner state shifted during the visit from feelings of 'normal' to feeling very emotional. When visitors were asked how they would describe the site in three words, the most commonly used words were 'terrifying, degrading, inhumane, oppressive and frightening'. These are very strong emotions and are related to the overall aura of the site, which was particularly visible during the visitor observations. I often witnessed how visitors paused as they entered the memorial site where one is immediately plunged into the prison atmosphere. In this regard, Bautzen stood out from the other three memorial sites where visitors did not use such strong words like 'inhumane' or 'degrading'.

The 'tiger cages' and the *Hörgang* (an area of solitary confinement with padded cells so that inmates could not hear any sounds) were the most distressing parts of the site, after which visitors often decided to leave. I accompanied a couple from Hamburg (but the female visitor from the former East Germany) whose visit was interrupted by another visitor who seemed in emotional distress. It transpired that he had been imprisoned at Bautzen, spending long periods of time in the 'tiger cages'. His psychiatrist had recommended that he confronts the site, yet it was obvious that he could not cope. This chance encounter left a mark on the couple I accompanied. They continued to visit the site but when they saw the 'tiger cages', the man said: 'okay, that's it. I'm going now. I've had enough', although emphasising that for him having an emotional experience at a memorial site was very important. The woman remembered at the beginning of the visit how her friend's flat was used by the Stasi to spy on a neighbour, and she explained that she did not really want to come here.

A couple from Forst/Brandenburg initially spent a significant amount of time on the ground floor, which details Bautzen's history and the work of the Stasi. When they encountered the *Hörgang*, the woman pulled a face, and she finally commented: 'I feel really strange now'. They were shocked about the tiger cages and could not believe that they did not have trained doctors at the prison until 1986. They continued to read some of the personal stories before leaving the site, when the man remarked: 'I'm glad I'm out of here now', and his wife added: 'I wouldn't want to have been in here.'

A group of visitors from North Rhine-Westphalia were equally shocked about the tiger cages and the solitary confinement area. They also repeatedly referred to the Bodo Strehlow story, saying: 'That's so awful.' Bodo Strehlow attempted to steer an NVA [GDR army] military ship into West German waters during his military service and was consequently sentenced to lifelong imprisonment in solitary confinement. He spent 10 years at Bautzen and was released shortly after the GDR's collapse in 1989. One of the men explained to me that the first time he went to Bautzen II he had to leave immediately as he could not deal with it. One of the younger visitors in the group thought that 'it is like the Nazi period, just in a different package', while another female visitor struggled to understand the history of the Soviet Special Camp. She said: 'Oh my God, they also had cells for four people', without realising that this cell represented Bautzen's Soviet Special Camp. When visiting the tiger cages she remarked: 'Any human dignity was taken away from them', to which her husband responded: 'when you imagine the psychological terror'. As she later visited the *Hörgang*, she said in shock that 'the headphones are extreme, you'll go mad listening to them'. At the end of the visit, they said: 'It is extreme what they have done to people. I'm wondering what the staff must have been like to behave in this manner.' This group, however, also reveals the 'Western' view of GDR history. At times, they behaved in a sensationalist manner by comparing it to popular prisons like Alcatraz. Moreover, comparing the GDR to the Nazi regime lacks understanding of the wider GDR history, a perception the visit to the memorial did not shift. In fact, the visit appeared to reinforce the GDR stereotypes. However, it was nevertheless an emotional visit, with one male group member commenting: 'I've never been in a prison and I don't think I'll go again.'

Not only the building itself but also typical GDR features such as the linoleum flooring evoked uncomfortable memories. A family from Bavaria immediately recognised the smell of the prison and said 'that is like the *Interzonenzüge* [trains between the GDR and the FRG] in the past'. They read individual stories before continuing to the *Hörgang* where the woman was clearly distressed. They then visited the outside space where the man commented, 'it also smells strange'. The woman emphasised at the end that she did not want to take part in 'scary tourism': 'I feel really ashamed. It is awful, I can go home afterwards. It is really strange to walk around here.' A couple in their 20s from the former East Germany also referred to the linoleum flooring immediately on arrival at the site. They also wondered what the sounds must have been like in the prison, and how visitors coped with this atmosphere. In fact, these visitors explained that they only know the GDR from hearsay since they have grown up in families with a direct experience of GDR life. For them, the GDR is a phantom that is present but also absent. And, despite not having grown up in the GDR, they recognised typical GDR features such as the linoleum flooring, a memory they adopted from their families. These experiences reveal that visitors are acutely aware of the emotional distress former prisoners experienced. In fact, they even appear to imagine themselves in the shoes of the prisoners, which leads to emotional contagion, thus crossing the boundary between them and the victim. Moreover, when visitors refer to the smell or sounds in a prison, they connect to the site in a sensual way. Thus, 'visits to museums or places of conflict pull on the imagination, on memory and emotion' (Palmer 2018, section 9) to transport the tourist back to a time when the events occurred.

Emotions were running high when visitors encountered perpetrators at Bautzen II. The management team decided to exhibit the biographies of the three prison managers. Two of them are still alive; in fact, one of them still lives in Bautzen. According to the biographies both receive a pension. This sparked outrage amongst visitors. One visitor from Hamburg could not understand why these managers received their pension while they 'lived happily' in Dresden and Bautzen. Referring to the former officer who still lives in Bautzen, he asked: 'How does he cope with what he has done?' He was subsequently shocked about the low conviction rate. Similarly, a couple from Hanover who had left the exhibition prematurely emphasised that '[i]t's really bad that these people [Stasi officers] were never convicted'. One visitor was so angry that he suggested that 'Honecker and Mielke should have been hanged'. Another male visitor from West Germany was incredulous at the fact that many former prison officers still work in Germany's prison service today, and asked: 'Has anybody asked them what they did in here?' There was no empathy for the prison staff apart from one West German male visitor who suggested that 'the staff must have been in shock when they arrived here'. It is not surprising that visitors responded in this way. Germany's GDR memory culture has, for a long time, placed its emphasis on the Stasi and its crimes. Films like *The Lives of Others* have added to this overwhelming perception of the GDR as an unlawful police state. Susanne Buckley-Zistel (2014), for instance, notes that after the screening of the film, visitor numbers at Hohenschönhausen skyrocketed. Bautzen II's victim-centred focus exhibition reiterates the notion of the stereotypical, ruthless Stasi officers. Although empathy with the victim is a key aim for Bautzen, not all visitors responded with this desired emotional state.

4.3.3 The instability of empathy

A visit to a memorial site does not necessarily lead to empathy with the victim; it can also result in 'empty empathy', a term introduced by Ann Kaplan (2011). Kaplan analysed the response to persistent exposure to traumatic images on TV which, if repeated in quick succession, can result in a shallow emotional response. Walking through a memorial site also involves coping with a frequent sequence of 'traumatic' scenes. Such scenes can overlay each other, thus not allowing the visitor to become too emotionally involved. In addition, an individual's own traumatic memories which can be triggered by being confronted with these violent images can prohibit feelings of empathy with other people's suffering.

For instance, a couple from Oranienburg/Brandenburg mainly focused on the personal stories and engaged little with the rest of the exhibition at Bautzen II. The man was sarcastic throughout and commented outside that 'it was not that bad'. In the *Viermannzelle* (Soviet Special Camp) he remarked that 'they still had wooden flooring', without realising that this cell was a reconstruction. At the end of the visit, he said:

I thought it would be much worse. I ended up in an NVA [Nationale Volksarmee, GDR army] prison and I wasn't even allowed to sit down. They had big windows here with lots of light. The awful atmosphere in the GDR with the grey facades can't be exhibited anymore anyway.

Similarly, a single male visitor from the Baltic Sea region seemed on the surface to engage extensively with the exhibition. He was particularly interested in the situation of the staff and read reports from former Stasi informants. However, after the visit he concluded that 'this is nothing here, I've been to the KGB prison in Potsdam, that is much worse'.

One family with three children from Baden-Württemberg, but originally from Saxony-Anhalt (East Germany), claimed that they wanted to educate their children. They visited the entire memorial site of Bautzen II (apart from the basement) including the outer spaces. Throughout the visit, the male visitor photographed the site, even barbed wire fences and GDR waste bins. He had an extensive knowledge of GDR history and commented occasionally that '[i]t is strange, one could walk around here and meet a former prisoner or a guard without knowing'. Outside he said: 'There would have been no grass, don't be fooled.' At the end of the visit he dispassionately commented that

> People nowadays cannot cope with anything, they are way too soft. I have seen much worse in the NVA. I know lots about this period, but one can only learn a lot at the authentic site. At the concentration camps, there aren't any eyewitnesses anymore but here there still are and that's why it has an edge.

He emphasised numerous times that he is not a voyeur and the only reason for the visit was to educate his children, yet he barely engaged with them.

At times, 'empty empathy' was intermixed with a sense of superiority and sensationalism. One single male visitor from Dortmund just wanted to know what a prison looked like and frequently compared Bautzen II to Alcatraz. He seemed slightly shocked about the tiger cages, but the positive childhood memories of visiting his grandparents in the GDR prevented any deeper understanding. Equally, another single male visitor from Gladbeck merely wanted to have a 'look behind the scenes'. 'I went to the GDR a lot to visit my grandfather in Dessau; I have great memories of my summer holidays.' He too was only interested in the prison itself, took a lot of photographs and also referred to Alcatraz occasionally. He acknowledged that 'the tiger cages must have driven you mad', but otherwise showed little empathy for the prisoners themselves and did not engage with the wider content. Another male visitor from North Germany walked around the building and asked me frequently 'whether there was something else interesting to see'. He briefly remarked at the tiger cages that 'that is degrading'. Equally, a male visitor from Düsseldorf was predominantly interested in taking photographs and, referring to one of the cells, rather sarcastically commented: 'This looks nice here'! He was also obsessed with the perpetrator and proudly announced that he had once sat at the desk of Mielke [head of GDR state security] at the Normannenstraße [headquarters GDR state security]. And finally he commented: 'I never read anything, I can read this at home on the Internet. I want to have a feel for the place. I have never been in a prison, I've only ever seen one on TV.'

These last four cases were visitors from the former West Germany who are emotionally distanced from GDR history. Emulating the experience of sitting at Mielke's desk was evidently more important than understanding the GDR dictatorship. There was also no awareness that short joyful summer holidays cannot be compared to living permanently under a dictatorship. The exhibition at Bautzen II did little to shift these perceptions. This affirms Crețan et al.'s (2019) findings in Romania, which highlighted that visitors at a memorial site do not necessarily reflect critically on what is presented to them. In Bautzen's case, visitors who arrived with a Western narrative of the GDR described earlier did not revisit this perception after and/or during the visit. Thus, a visitor needs to be prepared to engage with the content of an exhibition cognitively and emotionally, otherwise it is merely a museum visit like any other. Moreover, as Anna Kaminsky (2020) argues, the GDR is still viewed as a regional history that largely concerns the East Germans. In fact, several visitors from the former West Germany commented that they never learned anything about the GDR in school, and since they did not live near the inner-German border, they did not need to engage with this country. However, the lack of empathy also occurred amongst visitors who had a direct connection to the GDR. Here, their own experiences of having been in an NVA prison, perceived to be much worse, prohibited compassion with the prisoners' experience as their own suffering was viewed as exceptional. Nevertheless, Bautzen could also function as a space for dealing with one's own memories of the GDR, and not all visitors reacted in this dispassionate way as the following experiences show.

4.3.4 Personal connections

At Bautzen II some visitors either were victims themselves or had witnessed other people becoming victims. Much like the concentration camp memorials, the visit to the Stasi prison was a form of coming to terms with the GDR regime, in these cases experienced first-hand. A visitor from West Berlin, but born in the GDR, recalled the harassment he was subjected to as a young boy as both parents were political prisoners: his father was in Brandenburg for eight years and his mother was in Hoheneck for one year. He was subsequently not allowed to study at university. When he requested his own emigration from the GDR in 1976, he was then exposed to further harassment by losing his job. Within the exhibition, he was consequently most interested in documents about compensation and rehabilitation of convictions. Hence, the visit was a form of dealing with his own life experiences in the GDR. Similarly, two visitors from Hoyerswerda (close to Bautzen), a daughter with her elderly father, had already been once before to the memorial. They went briefly to the first floor and the basement when the father recalled: 'This was a terrible regime. I saw as a six-year-old how a man was arrested in the middle of the road; nobody ever knew what happened to him. The men in here must have suffered.' Here again, the visit is the framework for dealing with a traumatic childhood experience. As a young boy, he would have been unable to process the experience as he could not have made sense of it, so by confronting the memorial he tried to understand the fate of the arrested man.

The visit can also serve as a method of dealing with 'being on the other side' as the case of a male visitor who had worked for the West German consulate in the GDR demonstrated. As an employee he suddenly found himself in a position where he had to care for West German prisoners within the East German prison system. He said when he arrived at the memorial: 'I used to dread the drive from Berlin to Bautzen.' He had only ever seen the visitor's room and now needed to see the prison cells for himself. Subsequently, he recognised the signature of one of his colleagues on one of the documents. The personal destinies of the West German prisoners had never left him, in particular the story of a young woman whose one and only relative had died while, she, the prisoner, was serving a prison sentence. He recalled: 'I did not know how to tell her this. I decided to give her a hug and pre-warned the Stasi officers who were always present. They allowed me to do it, so at times there was some humanity.' He later was annoyed about the lack of information at the memorial site about West German prisoners; indeed they are not mentioned and the prison cells on the upper floors are not accessible: 'It does not state that there were West German prisoners here; this is so important!' The emotional impact these visits at the Stasi prison Bautzen II had on him, usually with little or no preparation, was visible. (He emphasised that he had no experience and/or formal training in social work).

Visitors also used the site to establish what happened either to themselves or to relatives. A single German visitor from Dresden, for instance, recalled being arrested during the demonstration in 1989. He was subsequently transported to an unknown place and only realised on arrival that he was in a prison in Bautzen. His visit was characterised by a desire to investigate the prison (Bautzen I or Bautzen II) he had been in. In the course of his brief visit it turned out that he had been in Bautzen I, which meant for him a proof of his own experiences. Equally a single man from Senftenberg/Brandenburg tried to establish what had happened to his stepfather and he immediately consulted the death book of the Soviet Special Camp. He realised later that his stepfather was imprisoned in Bautzen I, but not during the Soviet Special Camp era from 1945 to 1956. Hence, his stepfather might have been a GDR political prisoner in Bautzen I.

4.3.5 Competing historical narratives

One of Bautzen's difficulties is the incorporation of two very different historical periods within one confined space: one of them, the Stasi narrative, generally known, the other, the Soviet Special Camps, often completely unknown. In fact, since the summer of 2020, another layer was added: imprisonment under the Nazi regime. In a very confined space, Bautzen now addresses different victim groups that can easily lead to visitor confusion, a fact that the management team is aware of, yet it does not know how to rectify. Visitors' difficulties in locating the history of the Soviet Special Camps was certainly visible during my research.

For instance, one family from Bavaria was puzzled as they encountered the exhibition of the Soviet Special Camp and the reconstructed cell, with one of them subsequently commenting: 'If you [referring to me] had not explained this to me, I would have not understood it.' Equally, as he read the history of the Soviet Special Camp, another male visitor noted: 'I thought nobody had died here.' Although the Soviet Special Camp is explained in the exhibition, seeing both historical layers in close proximity to each other is hard for visitors to distinguish.

While the exhibition about the Nazi imprisonment did not exist during my visitor research, it is likely to add to visitors' confusion. As the visitor enters Bautzen II now, s/he is firstly confronted with an exhibition in the former Stasi offices. S/he then encounters an exhibition on the history of Bautzen I and II before entering the exhibition about the Nazi regime. In essence, the visitor moves between the Nazi period, Stalinism and the GDR within the space of a couple of minutes. Thus, Bautzen II is the perfect example of Michel Foucault's (2008) heterotopia which are spaces that have 'the power to juxtapose in a single real place several spaces, several emplacements that are in themselves incompatible' (19). As such, one cannot necessarily blame the visitor for being confused about this time capsule.

4.4 Summary

In this chapter I have shown how diverse the visitor experiences at the memorial sites are. For German visitors, it can be a balancing act of accepting Germany's responsibility for the Nazi past while also, at times, coming to terms with one's traumatic family memories and/or the family's involvement with the Nazi regime. Increasingly, however, the Nazi past is overshadowed by contemporary media representations and popular culture, which creates a conflict with Germany's scientific approach to exhibition design. Frequently visitors mentioned that they desire an emotional experience which is diametrically opposed to the aims of the German memorial sites. With an increased distance from the Nazi past, this will become an ever more pertinent issue for the memorial sites that commemorate the Nazi past. At the GDR memorial site, the research revealed the differences between the visitors for whom the GDR is a personal past and those for whom the GDR is largely a media spectacle. Thus, it is essential not to treat the GDR past as an East German concern as a lack of understanding of the GDR in the former West Germany might entrench existing stereotypes and misunderstandings.

4.5 Bibliography

Assmann, Aleida, and Juliane Brauer. 2011. 'Bilder, Gefühle, Erwartungen. Über die Emotionale Dimension von Gedenkstätten und den Umgang von Jugendlichen mit dem Holocaust'. *Geschichte und Gesellschaft* 37 (1): 72–103.

Bal, Mieke. 2010. 'Guest Column: Exhibition Practices'. *PMLA* 125 (1): 9–23.

Bar-On, Dan. 1989. *Legacy of Silence: Encounters with Children of the Third Reich*. London; Cambridge, MA: Harvard University Press.

Bogue, Nicole. 2016. 'The Concentration Camp Brothels in Memory'. *Holocaust Studies* 22 (2–3): 208–227.

Buckley-Zistel, Susanne. 2014. 'Detained in the Memorial Hohenschönhausen: Heterotopias, Narratives and Transitions from the Stasi Past in Germany'. In *Memorials in Transitions*, edited by Buckley-Zistel and Susanne Schäfer, 97–124. Cambridge, UK: Intersentia.

Buckley-Zistel, Susanne, and Susanne Schäfer. 2014. 'Memorials in Times of Transition'. In *Memorials in Times of Transition*, edited by Susanne Buckley-Zistel and Susanne Schäfer, 1–28. Cambridge, UK: Intersentia.

Cole, Tim. 1999. *Selling the Holocaust: From Auschwitz to Schindler: How History Is Bought, Packaged, and Sold*. New York: Psychology Press.

Crane, Susan A. 2008. 'Choosing Not to Look: Representation, Repatriation, and Holocaust Atrocity Photography'. *History and Theory* 47 (3): 309–330. https://doi.org/10.1111/j.1468-2303.2008.00457.x.

Crețan, Remus, Duncan Light, Steven Richards, and Andreea-Mihaela Dunca. 2019. 'Encountering the Victims of Romanian Communism: Young People and Empathy in a Memorial Museum'. *Eurasian Geography and Economics* 59 (5–6): 632–656.

Dalton, Derek. 2019. *Encountering Nazi Tourism Sites*. London; New York: Routledge.

Dudley, Sandra H. 2010. 'Museum Materialities: Objects, Sense and Feeling'. In *Museum Materialities: Objects, Engagement, Interpretations*, edited by Sandra H. Dudley, 1–18. London; New York: Routledge.

Duindam, David. 2017. *Fragments of the Holocaust: The Amsterdam Hollandsche Schouwburg as a Site of Memory: Fragments of the Holocaust*. Amsterdam: Amsterdam University Press.

Ebbrecht, Tobias. 2010. 'Migrating Images: Iconic Images of the Holocaust and the Representation of War in Popular Film'. *Shofar: An Interdisciplinary Journal of Jewish Studies* 28 (4): 86–103. https://doi.org/10.1353/sho.2010.0023.

Eschebach, Insa. 2011. 'Soil, Ashes, Commemoration: Processes of Sacralization at the Former Ravensbrück Concentration Camp'. *History and Memory* 23 (1): 131–156.

———. 2016. 'Einführung'. In *Von Mahnstätten über Zeithistorische Museen zu Orten des Massentourismus: Gedenkstätten an Orten von NS-Verbrechen in Polen und Deutschland*, edited by Enrico Heitzer, Günter Morsch, Robert Traba, and Katarzyna Woniak, 25–27. Berlin: Metropol-Verlag.

Faulenbach, Bernd. 1998. 'Zum Bildungsauftrag von Gedenkstätten in Ost- und Westdeutschland angesichts zweier Vergangenheiten und unübersichtlicher Geschichtsdebatten'. In *Bilden und Gedenken. Erwachsenenbildung in Gedenkstätten und an Gedächtnisorten*, edited by Heidi Behrens-Cobet, 23–34. Essen: Klartext-Verlag.

Ferrándiz, Francisco. 2020. *Contemporary Ethnographies: Moorings, Methods, and Keys for the Future*. New York: Routledge.

Finkelstein, Norman. 2014. *The Holocaust Industry: Reflections on the Exploitation of Jewish Suffering*. London: Verso Books.

Foucault, Michel. 2008. 'Of Other Spaces'. In *Heterotopia and the City: Public Space in a Postcivil Society*, edited and translated by Michiel Dehaene and Lieven De Cauter, 13–29. London; New York: Routledge.

Frevert, Ute. 2013. *Emotions in History: Lost and Found: Emotions in History: Lost and Found*. The Natalie Zemon Davis Annual Lecture Series. Budapest: Central European University Press.

Gryglewski, Elke. 2016. Personal conversation. *Exhibition Design House of the Wannsee Conference*, 25 August 2016.

Gudehus, Christian. 2004. 'Methodische Überlegungen zu einer Wirkungsforschung in Gedenkstätten'. In *Lagersystem und Repräsentation. Interdisziplinäre Studien zur Geschichte der Konzentrationslager*, edited by Ralf Gabriel, Elissa Mailänder-Koslov, Monika Neuhofer, and Else Rieger, 206–219. Tübingen: Edition Diskord.

Heyl, Matthias. 2016. 'Mit Überwältigendem überwältigen? Emotionalität und Kontroversität in der historisch-politischen Bildung. Ein Plädoyer für die Schärfung des Profils historischer Bildung'. In *Politische Bildung auf schwierigem Terrain: Rechtsextremismus, Gedenkstättenarbeit, DDR-Aufarbeitung und der Beutelsbacher Konsens*, edited by Jochen Schmidt, Steffen Schoon, and Landeszentrale für Politische Bildung Mecklenburg-Vorpommern, 37–55. Schwerin: Landeszentrale für Politische Bildung Mecklenburg-Vorpommern.

Jacobs, Janet. 2010. *Memorializing the Holocaust: Gender, Genocide and Collective Memory*. London; New York: I B Tauris & Co Ltd.

Jones, Sara. 2014. *The Media of Testimony : Remembering the East German Stasi in the Berlin Republic*. Basingstoke: Palgrave MacMillan.

Kaminsky, Anna. 2020. 'In der Mitte der Gesellschaft angekommen? Die Auseinandersetzung mit der Kommunistischen Diktatur in der SBZ und DDR im vereinigten Deutschland'. In *Erinnerungs- und Gedenkorte im sächsischen Dreiländereck Polen-Tschechien-Deutschland*, 65–91. Dresden: Sächsische Landeszentrale für politische Bildung/Umweltbibliothek Großhennersdorf e.V.

Kaplan, E. Ann. 2011. 'Empathy and Trauma Culture: Imaging Catastrophe'. In *Empathy: Philosophical and Psychological Perspectives*, edited by Amy Coplan and Peter Goldie, 255–277. Oxford; New York: Oxford University Press.

Keats, Patrice A. 2005. 'Vicarious Witnessing in European Concentration Camps: Imagining the Trauma of Another'. *Traumatology* 11 (3): 172–183.

Knigge, Volkhard. 2004. 'Museum oder Schädelstätte? Gedenkstätten als multiple Institutionen'. In *Museumsfragen. Gedenkstätten und Besucherforschung*, edited by Stiftung Haus der Geschichte der Bundesrepublik Deutschland, 17–33. Bonn: Stiftung Haus der Geschichte der Bundesrepublik Deutschland.

———. 2010. 'Zukunft der Erinnerung'. *Aus Politik und Zeitgeschichte* 25/26: 11–16.

Knoch, Habbo. 2020. *Geschichte in Gedenkstätten: Theorie – Praxis – Berufsfelder*. Tübingen: UTB Verlag.

Kosselleck, Reinhart. 2002. 'Formen und Traditionen des negatives Gedächtnisses'. In *Verbrechen erinnern*, edited by Volkhard Knigge and Norbert Frei, 21–32. München: C.H. Beck.

Kuchler, Christian. 2021. *Lernort Auschwitz: Geschichte und Rezeption schulischer Gedenkstättenfahrten 1980–2019*. Göttingen: Wallstein Verlag.

Kühling, Gerd, and Hans-Christian Jasch. 2017. '"Wer hier weint, hört nicht mehr auf". Zum Umgang mit der Wannsee-Konferenz und ihrem historischen Ort'. *Zeitgeschichte Online*. https://zeitgeschichte-online.de/kommentar/wer-hier-weint-hoert-nicht-mehr-auf.

Landsberg, Alison. 2004. *Prosthetic Memory: The Transformation of American Remembrance in the Age of Mass Culture*. New York: Columbia University Press.

Leo, Annette. 2000. 'Das Problem der nationalsozialistischen Vergangenheit'. In *Zweierlei Geschichte. Lebensgeschichte und Geschichtsbewußtsein von Arbeitnehmern in West- und Ostdeutschland*, edited by Bernd Faulenbach, Annette Leo, and Klaus Weberskirch, 300–347. Essen: Klartext-Verlag.

Light, Duncan, Remus Creţan, and Andreea-Mihaela Dunca. 2020. 'Transitional Justice and the Political "Work" of Domestic Tourism'. *Current Issues in Tourism*. https://doi.org/10.1080/13683500.2020.1763268.

Maier, Charles S. 2002. 'Die "Aura" Buchenwald'. In *Verbrechen Erinnern: Die Auseinandersetzung mit Holocaust und Völkermord*, edited by Volkhard Knigge and Norbert Frei, 327–342. München: C.H. Beck.

Manning, Jody Russell. 2010. 'The Palimpsest of Memory: Auschwitz and Oświęcim'. *Holocaust Studies* 16 (1–2): 229–256.

Meinhold, Philip. 2015. *Erben der Erinnerung: Ein Familienausflug nach Auschwitz*. Berlin: Verbrecher Verlag.

Möller, Lena. 2019. *'Auf Stätten des Leids Heime des Glücks': Die Siedlung am Vogelherd auf dem Areal des ehemaligen KZ Flossenbürg und ihre Emotionalisierung als Wohn- und Gedächtnisort*. Münster: Waxmann Verlag GmbH.

Morsch, Günter. 2018. 'Sachsenhausen Concentration Camp: Anniversary of Liberation22.04.2018'. www.dw.com/en/sachsenhausen-concentration-camp-anniversary-of-liberation/a-43483448.

Palmer, Catherine. 2018. *Being and Dwelling through Tourism: An Anthropological Perspective*. Abingdon, NY: Routledge.

Paver, Chloe E.M. 2018. *Exhibiting the Nazi Past: Museum Objects between the Material and the Immaterial*. Cham: Springer.

Pearce, Caroline. 2011. 'Visualising "Everyday" Evil: The Representation of Nazi Perpetrators in German Memorial Sites'. *Holocaust Studies* 17 (2–3): 233–260.

Rickly-Boyd, Jillian M. 2013. 'Existential Authenticity: Place Matters'. *Tourism Geographies* 15 (4): 680–686.

Sather-Wagstaff, Joy. 2016. *Heritage That Hurts: Tourists in the Memoryscapes of September 11*. London: Routledge.

Schilling, Erik. 2020. *Authentizität: Karriere einer Sehnsucht*. München: C.H. Beck.

Sereny, Gitta. 2001. *The German Trauma: Experiences and Reflections, 1938–2000*. London: Penguin.

Skriebeleit, Jörg. 2016. Personal conversation. *The Management of Flossenbürg*, 25 June 2016.

Stier, Oren Baruch. 2015. *Holocaust Icons: Symbolizing the Shoah in History and Memory*. New Brunswick, NJ: Rutgers University Press.

Tilden, Freeman. 1977. *Interpreting Our Heritage*. Chapel Hill: University of North Carolina Press.

Tyndall, Andrea. 2004. 'Memory, Authenticity and Replication of the Shoah in Museums: Defensive Tools of the Nation'. In *Re-Presenting the Shoah for the 21st Century*, edited by Ronit Lentin. New York: Berghahn Books.

Waxman, Zoë. 2017. *Women in the Holocaust: A Feminist History*. Oxford: Oxford University Press.

Witcomb, Andrea. 2010. 'Remembering the Dead by Affecting the Living: The Case of a Miniature Model of Treblinka'. In *Museum Materialities: Objects, Engagements, Interpretations*, edited by S.H. Dudley, 39–52. London; New York: Routledge.

Yair, Gad. 2014. 'Neutrality, Objectivity, and Dissociation: Cultural Trauma and Educational Messages in German Holocaust Memorial Sites and Documentation Centers'. *Holocaust and Genocide Studies* 28 (3): 482–509.

Zelizer, Barbie. 1998. *Remembering to Forget: Holocaust Memory through the Camera's Eye*. Chicago: University of Chicago Press.

5 German memory culture and tourism

Academic research into tourism ranges from a pure focus on business aspects (e.g. marketing or destination management) to anthropological perspectives (e.g. the cultural impacts on host communities). In so doing, research tends to concentrate on either quantitative or qualitative data, but we rarely see a more holistic approach. Nevertheless, these different research approaches not only have given us a great insight into tourist motivation and behaviours but also have shown us the destructive forces of overtourism. More recently, scholars have highlighted the phenomenon of rising visitor numbers to sites of death and distraction, coining the term 'dark tourism'. Although the ambition of the field is to transcend scholarly boundaries, its focus remains on tourists' motivations or on analyses of the commercialisation of death sites. Undeniably, the term 'dark tourism' gained traction and is now widely used. It has, however, also contributed to the stereotyping of tourists. The word 'dark' immediately evokes connotations of 'sinister', 'macabre', 'unpleasant', and whilst dark tourism scholars are keen to stress that there is no such thing as a 'dark tourist' (Stone 2019, 6), the use of the word 'dark' tapped into existing prejudices about the 'shallow, uneducated tourist'.

In fact, the uneducated tourist is a *Leitmotiv* that runs through tourism research. The British academic Daniel Boorstin (1972) describes tourists in his book *The Image – A guide to pseudo-events in America* as 'cultural dopes' since they expect to remain in comfortable tourist environments. Even worse are Louis Turner and John Ash (1976) who compare tourists to mass hordes that destroy local cultures and traditions. These criticisms have one key aim: to distract from the fact that oneself is also a tourist who will, whether one likes it or not, contribute to the negative aspects of any tourism development. In relation to memorial sites this distinction continues, with Tim Cole (1999), for instance, claiming that visiting Auschwitz is the ultimate rubberneck experience. An article in *The Guardian* highlighted poignantly the persistent trope of the superficial tourist at Auschwitz: the children's author Liz Kessler (2021) expresses her disappointment about tourists waiting for her to move away so that they can take a selfie. It was, according to the author, 'hard to stomach the idea of such places being on the tourist trail'. Prior to her visit in Auschwitz she had toured across Europe in a caravan, visiting various memorial sites; yet she was not 'a tourist'. She further argued that these visitors would trample on the

DOI: 10.4324/9781003126836-5

memory of the victims by not feeling the same despair than she did. Thus, she judges other visitors' behaviour based on her point of view, which we would call 'projection' in the psychological sciences. In fact, I argue that the constant criticism of 'inappropriate' behaviour distracts from the real issue: perhaps the memorial space does not have the transformative power that we thought it had, and we have indeed failed to achieve the 'never again'. Moreover, those who constantly denounce tourists often dismiss the possibility of the tourist's critical engagement with the site from the outset (Lisle 2016).

The persistent criticism of tourism at memorial sites might also be a cover-up for another uncomfortable truth: the inevitable fading away of Holocaust memory once the last survivors are gone. In a race against time, we are now recording the voices of Holocaust survivors who will be able to talk *virtually* about their experiences to future generations. An article in the *Frankfurter Allgemeine* (Sattler 2020) branded the new era of Holocaust memory as 'the era of memory remix' since the audio files can be rearranged according to 'taste'. All these initiatives are designed in the spirit of 'never forget' and 'never again', yet the 'end user' (to use the technologically correct term), the visitor, is suspiciously absent in these developments. Especially in the German context, we moved seamlessly from a lack of knowledge about physical tourists to a lack of knowledge about virtual tourists.

While German historians are aware of the end of the *Zeitzeugen* (witnesses to the past) and feel uncomfortable about Germany's memory culture, they appear to be trapped in a vicious cycle of criticism: they complain about ritualised commemorative pathos formulae (Jureit and Schneider 2010) or call it *Wohlfühlerinnerungskultur* (comfortable remembrance culture) (Wagner 2017). On reflecting on his work at German memorial sites, the German historian Bernd Faulenbach remarked in 2019 that Germany's challenge will be the increasing number of (international) tourists who visit the sites, and efforts should be made to avoid trivialisation. Faulenbach immediately connects tourism to Boorstin's concept of 'low culture' and suggests that for these visitors additional pedagogical efforts are required to engender empathy and cognitive interest. Considering the lack of knowledge about visitors at German memorial sites, an automatic assumption is made that the tourist is neither empathetic nor educated. As such, the tourist requires education rather than engagement. There also appears to be a difference between the tourist who visited concentration camp memorials in the 1970s and 1980s compared to the tourist who is visiting today. Yet, strictly speaking, most individual visitors to memorial sites are tourists and have been tourists for a long time.

I argue that tourism and tourist behaviour mirror our society. The rising visitor numbers at Auschwitz and other memorial sites are not due to a fascination with the macabre (Stone 2018) or an increase in media attention (Lennon and Foley 2010) but are a response to a changed memory culture. With the fall of the Iron Curtain, previously inaccessible sites such as Auschwitz were suddenly within easy reach. Moreover, the access to archives in the former Eastern Europe exposed new historical knowledge which mounted the pressure on local stakeholders to provide memorial sites that acknowledge different victim groups. The end of communist dictatorships in the former Eastern Europe also exposed the crimes that were committed under those regimes, leading to further demands for memorialisation of their victims. At the same time, the Stockholm Declaration, passed in 2000 by 46 governments, emphasised that 'the unprecedented character of the Holocaust will always hold a universal meaning', and committed the signing member states to commemorate the Holocaust. Since then, the International Holocaust Remembrance Alliance 'promotes Holocaust remembrance, education and research'. At the European Union level, several resolutions were passed that cemented the memory of the Holocaust as a 'negative founding myth' (Sierp 2020). However, with the enlargement of the European Union in 2004, the Eastern European states pushed for a stronger focus on the communist victims, which resulted in a new European-wide 'Day of Remembrance for the victims of all totalitarian and authoritarian regimes' on 23 August each year. Since there is now a greater awareness of historical atrocities, the increase in visitor numbers at commemorative sites comes as no surprise.

Thus, instead of dividing visitors into 'good' and 'bad' tourists and chastising those who do not behave according to the norm, we should ask how public discourses influence the tourism behaviour at memorial sites. In fact, I argue that tourists mirror contemporary memory practices and therefore shine a light onto our memory culture. Insa Eschebach (2020), for instance, notes that in the early years there was little interest, even among victim associations, in retaining buildings which were not directly related to mass murder. Conservation consequently focused on crematoria and/or the remnants of gas chambers. In the 1980s, the focus shifted to the wider concentration camp infrastructure, thus enhancing the historical understanding of the camps. It is therefore not surprising that today's visitors focus on the remnants as they are influenced by the memory narratives they have grown up with. The contemporary memorial site of Ravensbrück is, however, no longer a site that 'worships death', it now encourages reflective historical understanding. Thus, the academic concept of dark tourism is difficult to transfer to the German memorial sites as they do not 'package up the dead as commodities for consumption' (Stone 2020, 2). It is therefore my aim for this chapter to widen the view of tourist behaviour through a closer look at specific aspects that encompass the visitor experience at German memorial sites: empathy, the senses, atmosphere, photography, performativity, social identity and memory narratives.

5.1 Memorials as transformative spaces

There are regular demands from German politicians across the political spectrum to include visits to concentration camp memorials as a compulsory part of the German school curriculum. Although memorial managers and pedagogical staff regularly warn against such measures, not least due to the GDR's compulsory programme of visits, the warnings are largely ignored (Kuchler

2021). The German public is overwhelmingly in support of compulsory visits, stemming from the popular notion that Germany's youth must 'learn the lessons' and 'never forget'.

There is a belief system in society (and not just in Germany) that a visit to a memorial site (in particular Auschwitz) is a profoundly emotional experience, and learning about the Holocaust cannot be limited to the classroom. The visit to a memorial site is also seen as a form of vaccination against current societal difficulties: for example, after a racist incidence, Chelsea football players were sent to Auschwitz for a 'vital' lesson. Although there are no longitudinal sociological studies into the long-term effects of a visit to memorial sites, the belief persists that a visit to memorial sites transforms a visitor into an active defender of democracy and human rights. Yet Sarah Gensburger and Sandrine Lefranc (2020) argue that the rise of antisemitism and racism suggests that Holocaust memory is not very effective. When one considers the rise of violent acts, an equally pessimistic picture emerges. Yet a survey of Holocaust memory across different nationalities confirms that most participants feel that the preservation of concentration camp memorials is essential to never forgetting. Indeed, during my exit interviews, visitors at all sites supported the belief that the memorial sites should remain open for future generations. An exception was one Israeli visitor (but born in Yemen) at the House of the Wannsee Conference who thought it was better to move on during my interview, which upset his Jewish wife. Thus, visitors overwhelmingly supported the notion of 'learning from the past', even though for most visitors the humanitarian values that memorial sites transmit are secondary (Gensburger and Lefranc 2020).

Christian Kuchler's (2021) research into student trips to Auschwitz between 1980 and 2019 also shows that these universal assumptions of 'learning the lessons' are misplaced. In the 1980s, West German pupils saw themselves confronted with a Polish exhibition design that overwhelmed them. Historical sources revealed that pupils often left the site in tears. Kuchler concludes that educational opportunities were very limited in those days. Crucially, most pupils recalled Holocaust relics (e.g. shoes and hair) rather than historical knowledge. Although the nature and style of these student journeys have changed significantly since then, Kuchler's research emphasises how exhibition design and educational understanding interact. Moreover, whether a memorial site has a transformative power also depends on the reflective opportunities that are provided throughout the visit.

Hartmut Rosa's (2019) concept of 'resonance' explains the crucial components for a transformative experience, which can elucidate the reason for some visitors are touched by the exhibitions, while others remain detached: firstly, it requires affection, which is an encounter that truly touches us, thus requiring an emotional, bodily and cognitive response; secondly, it needs to engender emotions that encourage us to reach out with body and mind; and thirdly, the process of being touched and responding emotionally must transform us. For instance, the visitor at Flossenbürg who cried when encountering the story of the young girl for whom liberation came too late was affected by this encounter. She later responded to it by reflecting quietly in the chapel at Flossenbürg. Thus, for a memorial site to be transformative, visitors need to be able to play an active part in engaging with the past that is presented to them. If the visit, however, is characterised by an overwhelming need for an authentic experience of an imagined past that presents itself in the form of watchtowers or barbed wire, then a transformation might not take place. Rosa (*ibid*) also highlights that resonance is elusive as it is impossible to know or to predict when it will occur and what the process of the transformation will be. Thus, in the context of the memorial sites, we cannot foresee or assume that each visit will result in visitors being transformed into ethical human beings. In addition, Rosa's theory of resonance reveals another tension at German memorial sites: an emotional response is not desired since the focus is on cognitive aspects of learning, yet emotions are essential for a transformative experience. Yet an analysis of Instagram posts by Iris Groschek (2020), Neuengamme's memorial manager, and my own research certainly revealed that there is a strong desire for an emotional experience at memorial sites.

5.2 Empathy

Bernd Faulenbach (2019), on reflecting on German memory culture since 1990, highlights the centrality of 'empathy' as a key outcome of a visit to a German memorial site, and not just in Germany. Yet we still do not exactly know what empathy is, and it is made worse by the interchangeable use of the terms 'empathy' and 'sympathy'. If we are unclear what empathy is, then it is hard to quantify as an outcome of a visit to a memorial site. Ute Frevert's (2013) research into the history of emotions revealed that 'empathy' is one of the emotions that is learned in society, as opposed to the belief in the psychological sciences that emotions are inherited.

Although *Mitfühlung* has a long history, it was not until the 18th century that the term was gradually given a moral dimension. The German encyclopaedia *Brockhaus*, for instance, placed *Mitempfindung* (empathy) above sympathy by defining it as 'the spontaneous imitation of someone else's sentiment' (cited in Frevert 2013, 176). Frevert (*ibid*) clarifies that empathy is connected to the German term *Einfühlung*, which requires us to enter another person's emotional state. Yet this involves a structural dilemma: firstly, we have to imagine ourselves in the shoes of the other person, and secondly, we have to assume that we know what the other person would have felt in a particular moment. If we transfer this concept to memorial sites, it requires tourists to imagine themselves in the shoes of the victim (or maybe the perpetrator) and to know what the victim (or the perpetrator) would have felt.

Yet the very idea that one can know what a victim would have felt at a concentration camp or in a Stasi prison is generally considered to be an inappropriate form of annihilation. Gary Weissman (2004) even considers it to be a fantasy as one can never know what the former prisoners felt. Furthermore, imagining oneself in the victim's shoes can also lead to emotional exhaustion or even trauma, as the research by Michał Bilewicz and Adrian Wojcik (2018) with high school students in Poland has shown. This research has revealed that visitors commonly used the German term *mitfühlend* when describing their emotional state after the visit, which leads me to conclude that visitors tend to describe the culturally expected emotion rather than their true feelings. In fact, Kuchler's (2021) research into the history of school trips to Auschwitz found that it was almost impossible to understand the true nature of students' emotions since they so often responded in the culturally appropriate way.

In addition, the ethnographic research at the memorial sites has highlighted that one's own traumatic memories can inhibit feelings towards the victim. Thus, in order to feel empathy for a victim, one has to possess a level of emotional maturity. A response during an exit interview at Bautzen II demonstrates this point: the visitor replied when asked about his emotions that he does not feel empathy towards the prisoners since they were aware that *Republikflucht* (fleeing the GDR) was a crime under GDR law; hence, they knew that they might face the risk of being imprisoned. Bautzen II was, however, also the site where the victim-centric exhibition often led to the desired empathetic response, which in turn could also result in feelings of not being able to cope, and therefore a premature termination of the visit. The strong empathy for the victim also sparked an emphatic condemnation of the prison officers, the perpetrator in the eyes of many visitors. In this regard, Bautzen's exhibition perpetuates the scandalisation of the Stasi and the black and white presentation of the GDR which took place since the fall of the Berlin Wall. Only one visitor I accompanied wondered about the feelings of the members of staff in the prison, as that they might have not worked in the Stasi prison out of choice.

The nature of empathy also seems to change with increasing distance from the historical events. Whilst visitors at the concentration camp memorials certainly showed signs of empathetic matching, this can be explained as vicarious emotions. Paulus et al. (2013) argue that an empathetic emotion requires social interaction with another person, while vicarious emotions can occur in the absence of another person's presence and therefore requires extensive context. When visitors shake while walking into a crematorium, they are aware of the locational context; without it, they would regard the crematorium like any other crematorium at a cemetery. One could, of course, argue that at a Stasi prison the other person is also absent, but one couple's encounter with a former prisoner resulted in an intensely emotional reaction, which supports the argument that empathy requires more than an exhibition. Indeed, this is in line with Crețan et al. (2019) whose findings concluded that, while a memorial can *plan for* an empathetic reaction, it cannot influence the outcome.

The question that remains is whether we should strive for empathy as an outcome of the visit. Although empathy contains an active part as it requires us to imagine ourselves in the position of the other person, it is also passive. We know that we do not have to act since we are thankfully not the other person. Ute Frevert (2013) notes that in the 18th century there was an awareness that feeling compassionate or sympathetic was not enough if it did not lead to an active involvement in fighting social injustices. Most visitors at all memorial sites claimed that the visit did not encourage them to review their own behaviour. In fact, I encountered visitors who held a negative view of Germany's then recent immigration policies. Thus, focusing extensively on empathy in exhibitions might not result in a critical engagement with the content. Indeed, it questions the commonly used narrative of memorial sites as transformative spaces.

5.3 Tourism and collective memory

Friedrich Porsdorf (2019, personal communication), a GDR artist at the *Akademie der Künste* in Berlin, was asked in the 1980s to get involved in the redesign of the exhibition at Ravensbrück. He subsequently painted scenes of the concentration camp onto walls for the newly developed exhibition in Ravensbrück's *Kommandantur*. During an interview with me, he recalled staying overnight at the former camp which haunted him and reminded him of the end of the Second World War in Radebeul (Dresden) where he witnessed prisoners in striped uniforms walking through the town. Friedrich said he was aware that the GDR's focus on the communist victims was incorrect, and he tried as much as possible to paint the 'true' experiences of the women in the camp. After the fall of the Berlin Wall, he was informed that his paintings would be removed since the new exhibition could no longer accommodate them. Although he did not like the proposal since it would inevitably destroy his artwork, he had to accept the management team's decision. After all, he said, 'GDR art was not worth much anymore'.

This interview highlights the challenges of coming to terms with a new narrative of the Nazi past after German reunification. Friedrich's story, however, also reveals that the GDR's antifascist narrative was not universally adopted. In fact, Annette Leo's (2000b) oral history project has shown that the generation born into the GDR was most likely to have adapted the GDR's narrative, while the first generation still had their own memories of the Second World War, and the third generation were already influenced by the new memory culture after German unification. Nevertheless, experiences of visiting the memorial sites in the GDR linger. A visitor from the former GDR shook her head when I replied that I would also conduct research at Ravensbrück after she had questioned me about my research. She recalled the gruesome images of the medical experiments in the exhibition and said she would never go to Ravensbrück again. Moreover, East German visitors at Ravensbrück remembered the more emotional exhibitions and challenging images, which is not surprising as Inga Kahlcke's (2017) analysis of GDR history schoolbooks highlighted that the GDR frequently showed shocking images. By contrast, visitors from the former West Germany were surprised to find a women's concentration camp on German soil. And a British couple remembered on arrival Odette Sansom Hallowes, who, as former Special Operations Executive (SOE), became a war hero in post-war Britain. These examples emphasise that collective memory narratives and tourism go hand in hand and have a significant influence on how a visitor responds to the site. Volkhard Knigge's (2004) observation that all visitors might be in Buchenwald but are not in the same place is accurate. What these examples also stress is that memorials sites are melting pots where several narratives meet: the family, the national and the global narrative intersect while visitors negotiate a path between them.

The influence of memory narratives was also visible at Bautzen II. One visitor from Tirschenreuth (located in Bavaria at the former inner German and Czech border), for instance, recalled how his father used to warn him about the dangerous GDR. Another visitor remembered Löwenthal's TV show on human rights violations in the GDR. The show's aim was to unravel the political situation in the Eastern bloc, and in particular in the GDR. Löwenthal also encouraged GDR citizens to send in letters which he would read out under the headline *Hilferufe von Drüben* (calls for help from the other side). The show itself was controversial in West Germany, often criticised by the Left for its negative judgment of Willy Brandt's *Entspannungspolitik*, yet supported by the right-wing political spectrum for its anti-communist stance. West German visitors will, therefore, be influenced by the decades of 'othering' of the GDR citizens when they arrive at Bautzen II. In addition, Annette Leo (2000a) points out, based on oral history interviews with East and West Germans in the 1990s, that there are three different West German groups with memories of the GDR: the refugees who predominantly remember oppression and despotism, the West German relatives who saw the GDR occasionally during holidays and particularly remember the unpleasant border crossings and the 'transit' tourists who recall empty shops and dilapidated houses. There is also a fourth group, those who never went to the GDR and had no direct contact with GDR citizens. Although the groups have different memories of the GDR, they have one perception in common: when they talk about the GDR, they refer to the machinery of power. In contrast, the former GDR citizens avoid speaking about repression and instead emphasise the *Lebenswelt* (everyday interactions and experiences). Thus, combined with the strong focus on the Stasi in the early years after German reunification, a visit to Bautzen II will add to the West German's preconceived ideas of what life in the GDR was like. Moreover, a younger visitor from the former West Germany who says that 'the GDR was like the Nazi regime just in another package' adopted the post-reunification narrative that frequently compared the GDR with the Third Reich (Berdahl 1999).

Hence, a visitor to a memorial site is also never 'apolitical'. Debbie Lisle (2016) argues that the dark tourism literature interprets the phenomenon entirely within the geopolitical logic. Enlightened (largely Western) communities travel to sites of atrocities gazing upon the misfortune of another society in the safe knowledge that their nation has overcome such struggles or never had to deal with them in the first place. In essence, the tourist is the custodian that will uphold democratic values (*ibid*). How this dynamic works in practice was visible at Bautzen II. When a 'West' German visitor refers to 'Mielke's desk in the Normannenstraße' s/he perpetuates the contemporary German narrative:

the enlightened West has superseded the failure of state communism. Daphne Berdahl (1999) described poignantly in her ethnographic research at the former border region in Thuringia how the sense of superiority led to increasing hostilities between the former East and West. East Germans were frequently belittled and described as lazy and backward. As a consequence, most East Germans retreated into a shell, burying their memories of life in the GDR in an attempt to fit into the new system. The discrepancy between the state narrative, which reduced the GDR to the dictatorship and the Stasi, and the individual's memory created an explosive mix that breaks open in Germany today (Arp 2021). In post-conflict societies tourism has therefore a significant political dimension that we cannot ignore. In Germany, journeys to Stasi prison memorials might reinforce stereotypes that can enhance existing hostilities.

5.4 Tourism and German memory culture

Kuchler (2021) points out in his research on the history of West German school trips to Auschwitz that in the early 1970s a visit to Auschwitz was just a stop amongst many others on a journey through Poland. Reports about the trips to Poland often did not specifically mention Auschwitz, which changed with West Germany's stance towards the Nazi past and in particular the Holocaust. With the broadcasting of the TV series *Holocaust* in 1979, Auschwitz was no longer suddenly only a stop on a busy itinerary; it became the place of significance in Germany's culture of remembrance. Consequently, memory discourses influence tourists' experiences. Today, Auschwitz is the global symbol of the Holocaust, with visitor numbers far exceeding its capacity, resulting in the German government advising on 'alternative sites' during school trips to Poland.

A pivotal moment for Germany's reckoning with the past was the '1968 generation'. This generation desperately attempted to break with the silence of their fathers by focusing on unearthing history, resulting in numerous grassroots movements to commemorate the Nazi past. The emphasis was on securing and recording 'the traces' of the Nazi past; exhibitions were often a by-product and designed with very little pedagogical intervention. Yet this generation focused on the structures of the Nazi regime rather than the individuals. Many German visitors at the House of the Wannsee Conference, Ravensbrück and Flossenbürg belonged to the generation who grew up surrounded by silence, often symbolised in the phrase 'I have learned nothing in school'. This generation of visitors has a different connection to the Nazi past. Whilst they might have not been involved in committing the atrocities, they grew up in the shadow of them. For these tourists, a visit to a memorial site can be a personal reckoning with the past.

Yet a new generation of German visitors no longer has this connection to this Nazi past. They are predominantly influenced by Germany's - and an increasingly global - memory of the Holocaust. They have now also grown up with the media images of the Holocaust and school trips to various memorial sites, including Auschwitz. Thus, a visit to a German memorial site that is characterised by absence does not touch them in the same way that the previous generation responded. So, whilst the generations have changed, Germany's approach of engagement with visitors at memorial sites has not. Moreover, Germany's memorial sites are increasingly visited by international visitors who arrive with their own memory culture. However, these tourists, according to Faulenbach (2019), need a specific education in order to understand the historical significance of the site. In essence, the tourist is expected to cope with and comprehend Germany's memory culture. Such approaches have already failed in Flossenbürg where most American visitors could not follow the 'What remains' exhibition which deals with German memory politics. In addition, it ignores Germany's increasingly diverse society which might view the Nazi past in a different light due to their own traumatic experiences.

A couple I accompanied in Ravensbrück complained about Muslim visitors taking photographs as they apparently could not comprehend the significance of the site. Germany's memory culture has fostered a style of remembrance that excludes sections of society. Esra Özyürek (2019), for instance, explains that since the early 2000s European newspapers commenced running stories about the lack of engagement of Muslim communities with the material culture of the Holocaust and/or avoiding trips to concentration camp memorials altogether. Thus, a moral superiority over the correct format of Holocaust remembrance developed which accused migrants of lacking the Holocaust knowledge to be able to behave appropriately. The couple who criticised the behaviour of the Muslim visitors had embraced this narrative. Although not comparable, the East Germans were another group that was marginalised in the Holocaust memory discourse. With the fall of the Berlin Wall, the former GDR citizens, who had grown up with the antifascist narrative, had to be educated about the true nature of the Nazi regime by changing the exhibitions that were based on West Germany's memorialisation model. Once the exhibitions were revised, it was up to the East Germans to 'catch up with, adapt to, and later simply adopt this system' (Berdahl 1999, 159). Yet as my research has shown, simply redesigning exhibitions does not automatically change 'memories'.

5.5 Identity formation

During an exit interview at Flossenbürg a visitor said to me: 'You cannot possibly be proud of being German; wherever you go in Europe the Germans have caused devastation.' This remark corresponds to the recent research into German identity conducted by the Bertelsmann Foundation (Bertelsmann Stiftung et al. 2020). The core finding was that the participants were unsure of what being German meant. In addition, the memory of the Nazi regime made most participants in East and West Germany feel uncomfortable expressing pride in Germany. By contrast, the fall of the Berlin Wall is judged differently: East Germans prefer the use of the word *Wende,* thus emphasising the political and personal upheaval, while West Germans refer frequently to the word *Wiedervereinigung* (reunification), therefore regarding it largely as a 'natural' political process. Thus, even 30 years after reunification, Germany is still somewhat divided and Germans do not experience a national pride. On the other hand, Germany's new identity as a nation-state, based on the responsibility and memory of the Holocaust, is successful.

Identity formation is a crucial aspect of tourism. On a collective level, tourism can be used to cement state narratives. It is, for instance, no surprise that the GDR actively promoted its concentration camp memorials in tourist guidebooks since it strengthened its foundation. On an individual level, identity in tourism can be viewed as an encounter with the 'other', which helps in self-actualisation processes and group-belonging. By travelling to certain destinations, I portray to the outside world who I am, thus identity is closely linked to lifestyle (Giddens 1991). The encounter with the 'other', however, also demands that I interrogate my identity. At memorial sites, Henry Tajfal and John Turner's (2004) concept of social identity comes to the fore as I have to position myself as belonging to the perpetrator, the victim or the bystander group. This positioning also depends on visitors' social *Erinnerungsmilieu* (the family narrative). For instance, a visitor whose family openly discussed perpetrators within the family will respond differently to being confronted with perpetrators at a former concentration camp than those whose family silenced the topic (Welzer et al. 2002). In addition, our social background informs our image of history which visitors will deepen, change or reject during a visit to a memorial site (Eckmann 2010). Such confrontations can be highly emotional and challenging for visitors.

One could observe the identity formation processes predominantly with German visitors. Some visitors acknowledged Germany's responsibility for the Second World War while at the same time distancing themselves from this past. A visitor from Saxony, for instance, proclaimed after her visit to Ravensbrück that 'it was awful what happened here but still Germany does not need to invite them [refugees]'. She clearly recognised Germany's humanitarian role, yet rejected her personal responsibility. The visitor at Flossenbürg who compared his childhood home to a concentration camp made a similar remark: 'Of course, I feel sorry for the refugees, but I have to work for them', meaning that his taxes will be spent on supporting refugees. One's own identity crisis is projected onto the other, less fortunate group: refugees. In fact, the comment 'I am annoyed that Germany is still held responsible for this past' was not uncommon at the concentration camp memorials.

Most striking, however, was the wrestling with one's family involvement in the Nazi past. In particular, the House of the Wannsee Conference has shown the challenges in dealing with family members who actively supported the Nazi regime and in some cases continue to do so. For subsequent generations acknowledging the Nazi past has become normalised in Germany, yet within families the same normalisation does not necessarily take place. For instance, the young woman at the House of the Wannsee Conference who remembered the loving grandmother with Swastikas on the wall explained that her father categorically refuses to talk about his parents' involvement with the Nazi regime. In this case, the complex family history was used as a foundation to actively deepen the knowledge about the Nazi regime, so much so that the exhibition became overwhelming for her.

At Bautzen II, several identity formation processes emerged. For the visitors from the former West Germany, Bautzen confirmed that they had lived in the 'better' Germany, in contrast with their cousins *drüben* (over there). For East Germans, however, it could be an uncomfortable reckoning with themselves. Many of them stressed that they had never had anything to do with the Stasi and they had avoided visiting the site for many years. Germany's GDR memory culture has focused on the Stasi which, according to the Bertelsmann Foundation's (2020) research, East Germans have absorbed into their own memory. On the other hand, they do not want to belong to a group that is solely defined by coming from an 'unlawful' state. I therefore interpret the excessive sarcasm some visitors displayed as a defence mechanism against troubling emotions that might confirm what one knew about the GDR but wanted to bury. The historian Ilko-Sascha Kowalczuk (2021) emphasised that the Stasi had an omnipresence in the GDR, which means that even though one might not have had direct contact, one was definitely aware of them. After German reunification many people forgot this very quickly, yet it led to an inability to trust others. A visit to a Stasi prison might stir up such long-forgotten emotions.

5.6 Sensual experiences and atmosphere

Comments such as 'I had shivers running down my spine' at the House of the Wannsee Conference reveal the significance of embodied experiences. And although pursuing such experiences is a key component, there is currently no sustained effort to analyse the concept of embodiment in tourism (Palmer and Andrews 2020). Reviewing the literature of memory studies or dark tourism studies reveals that the sensual experiences of visitors at memorial sites are almost entirely absent, with the exceptions of Shanti Sumartojo's (2016) work on Anzac Day and Camp de Milles in France and Bird et al.'s (2020) research into the tourists' experiences of the First World War battlefield sites.

Memorial landscapes and exhibitions are designed to evoke a reaction, such as the experiential path at Flossenbürg, designed in the way of the cross, or the walk from the dark to the light at Ravensbrück. A visitor walking through such a landscape is encouraged to connect to the site using the senses, yet it nevertheless requires imagination as the sites themselves are characterised by absence. Tourists, therefore, have to rely on their pre-existing knowledge about the site, so embodied reactions arise from within the visitor and are triggered by encountering remnants of the past. Thus, social memories play a crucial role in embodied reactions and explain why a German visitor feels emotionally moved at a concentration camp memorial, while a Swedish visitor enjoys the beautiful landscape. Although at the House of the Wannsee Conference visitors are not invited to embody the perpetrator, the (previous) exhibition attempted to tell the story of the perpetrator alongside the victims. Hence, when visitors talked of a 'chilly atmosphere', they re-enacted the fate of the victims which the conference discussed. One could therefore argue that Wannsee's concept of avoiding identification with the perpetrator was successful, but, on the other hand, by re-enacting the victim the visitor remains in a comfortable position as s/he will not contemplate of 'having a little Himmler inside' (LaCapra 1994).

Flossenbürg's and Ravensbrück's aesthetics encouraged performative acts even further. The ground floor of Flossenbürg's exhibition features the former baths and the 'hairdresser'. These rooms contain very few explanatory texts and are mostly left in their derelict state, although they have been somewhat renovated. Indeed, the 'shower room' still contains the shower heads and the tiles from the period of the concentration camp. Visitors I accompanied were often reluctant to enter the 'shower room' and most of them preferred to remain at the door, commenting that the space felt 'strange'. Here tourists associate the shower room with the gas chambers of extermination camps and therefore perceive the space as tainted. Similar embodied experiences could be observed at the crematorium which is located at the bottom of the 'valley of death' and surrounded by dense woodland. Even on a bright summer's day the location itself is dark. Thus, the cold atmosphere creates a sensual experience that triggers visitors' memories of the fate of the victims. Visiting the former factory at Ravensbrück required walking across the former area of the barracks, now covered with a black surface made of clinker, giving the illusion of a desolate space. I accompanied visitors who felt reluctant to walk across the surface, whilst for others walks across the memorial site brought back uncomfortable family memories. The act of 'being' at the memorial site can create deep personal connections as a response to its aesthetics and atmosphere. In fact, I argue that these sensual experiences are a key motivator for visiting the site.

During my research at Flossenbürg, I decided to explore the quarry in which prisoners were forced to work, but which is currently not part of the memorial site. The path to the quarry leads you through quiet woodlands before you suddenly stand in front of the huge granite blocks of the current onsite extraction, and the decaying former SS buildings. It was, however, not the quarry that made me feel uncomfortable, it was the eerie walk to the site. I had a similar experience at Ravensbrück where I explored the former youth protection camp Uckermark, which is a grassland today. At both locations, very little reminds the visitor of the previous trauma, and yet there was a sense of unease that is difficult to describe. Perhaps it was my historical knowledge about the sites and being completely on my own that caused this sense of dread. However, it was not only me who could at times 'feel an atmosphere', visitors frequently commented on the memorial site's atmosphere. Although atmosphere is difficult 'to put your finger on' (Turner and Peters 2015, 313), it is essential to take it into consideration since it is closely linked to the sensual experiences at the memorial sites.

Gernot Böhme (2017, 18) suggests that atmosphere is 'what is in-between' a subject and an object. Atmospheres are not 'things'; they cannot exist without the subject feeling them; as such, atmospheres are 'quasi-objects': you can enter them but you can also be caught by them by surprise. Böhme (*ibid*) further highlights that atmospheres are always spatial and always emotional. Consequently, atmospheres can be approached from a perception of aesthetics or production of aesthetics point of view, that is, atmospheres can be researched using people's descriptions of a site, but atmospheres can also be staged through architectural design. Interestingly, at memorial sites, these two viewpoints merge. Ravensbrück and Flossenbürg, with their experiential landscape designs, are 'staged atmospheres' in which nature, however, interferes by reclaiming the spaces.

Böhme views atmospheres as a product of aesthetics, which does not explain why visitors at the House of the Wannsee Conference refer to the house 'being chilly', considering that the management team's main aim is to 'break the dark spell of the building'. Therefore, Shanti Sumartojo and Sarah Pink (2018) argue that atmosphere is emergent and influenced by anticipation. For instance, the visitors on Anzac Day arrived in a sombre frame of mind, which influences how they respond to the site. In these cases, the atmosphere is not produced in the particular moment, it reaches back to previous places and experiences. For example, the visitor at the House of the Wannsee Conference who did no longer experience shivers during his second visit had lost the sense of anticipation. He now knew that the building does not resemble the conditions of the Wannsee Conference in 1942. Similarly, the visitors who referred to Auschwitz as being worse than Flossenbürg and Ravensbrück missed an atmosphere. They remembered how they felt when they were face-to-face with mass murder, which the German sites marked by absence cannot replicate.

Hence, there is a link between memory and atmosphere. Cues at memorial sites can trigger memories that contribute to the atmosphere. And whilst memorial managers and/or architects can influence the development of an atmosphere, they can never predict how and in what manner the visitor will 'enter the atmosphere'. It also highlights that the focus on representational strategies or discourses in museums and at heritage sites in memory studies omits the feelings, sensations and thoughts that people experience at these sites (Sumartojo and Pink 2018). When visitors respond that the hairs are standing up on their necks, then we can see that visitors understand historical places in atmospheric terms. Thus, concepts in memory studies that view memorial sites as culturally fixed (e.g. Assmann 2011) overlook the dynamics that emerge through interactions with landscapes, objects and bodies (Muzaini 2015).

Nevertheless, visitors must be aware of the markers at a memorial site for an atmosphere to arise (Drozdzewski et al. 2016), for instance the indication by white lines of the former *Sonderbauten* at Flossenbürg is a marker. If visitors then respond by shaking, then they become conscious of its historical significance, which in turn can lead to a sense of uneasiness. If, however, the site would be merely a grassland, then most visitors would not sense an 'atmosphere'. Hence, my own sense of dread during the walk to the quarry at Flossenbürg derived from the awareness of the historical markers, as I imagined that most prisoners would have used this path on their daily march to the quarry.

Bird et al. (2020) emphasise that the further in the past the historical event lies, the more difficult it is to create this emotional connection, which explains why visitors at Bautzen II had a greater sensual experience. Since Bautzen II's main history as a former Stasi prison is still in living memory, visitors noticed the smell of the linoleum flooring or pointed at the wallpaper in the director's office. Here memory, imagination and emotion converge (Palmer 2017), reminding some visitors of their former life in the GDR. Thus, the sensual experience is more intense than at the memorial sites of the Nazi past. Moreover, the aesthetics of the prison itself evokes primal fears – the loss of freedom. So even if one has no direct connection to the Stasi past, most visitors will be able to relate to the sense of powerlessness. Thus, some visitor reactions emerge as a response to the eerie nature of the site. Kimberley Peters and Jennifer Turner (2015), in analysing the Eastern State Penitentiary and the National Justice Gallery in Nottingham, stress that these prison museums do not just use artefacts to create a carceral atmosphere, they use the remnants that haunt these disused buildings to produce an 'uncanny sense of realism' (311). Whilst Bautzen II is not a crumbling building, it retained the typical GDR features. Museums do not make such decisions unintentionally; they are part of a meaning-making process to elicit a politically charged atmosphere. Whilst it is Bautzen's aspiration not to overburden the visitor, it is the preservation of the GDR features that gives rise to intense emotional reactions in response to the atmosphere.

5.7 Photography

Closely linked to the sensual experiences at memorial sites is the question of visitors taking photographs, and in particular 'selfies'. Self-appointed arbitrators (Clarke 2019) often condemn photography as a trivial and disrespectful activity. Although an analysis of images on Instagram at the memorial sites Dachau and Neuengamme has shown that visitors barely post images that contain a person, selfies at concentration camp memorials are regularly scandalised in the media (Groschek 2020). Questions about the type of photographs and about the person who takes them are rarely asked.

At Flossenbürg, Ravensbrück and Wannsee, German visitors rarely took photographs, and if they did, they tended to be panoramic shots rather than a focus on specific aspects of the site. In stark contrast to this were international visitors. At Flossenbürg popular motives were the watchtowers, the crematorium and the American memorial plaque. At the House of the Wannsee conference, the conference room (with a copy of the protocol), the view across the lake and the villa itself were also popular subjects. This highlights that German visitors appear to experience a cultural barrier that prevents them from taking photographs. Thus, taking photographs at memorial sites is a performative act. Pierre Bourdieu (1990, 75) argues that the ordinary photographer constructs the world as s/he sees it. Analysing photographs at memorial sites, therefore, requires an analysis of the frame:, what/who is (or is not) in the frame. When American visitors at Flossenbürg photograph the watchtowers or the ramp that connected the crematorium with the upper part of the concentration camp, then they construct *their* notion of a concentration camp. Photography is consequently a form of identity construction as it exposes my relationship to the past. Moreover, when concentration camp memorials are reduced to barbed wire or watchtowers in photographs, it uncovers the underlying memory culture and the striving for authenticity.

German visitors as members of the former 'perpetrator society' do not want to be part of the 'frame'. For these visitors, the Nazi past is *their* past and one of profound shame. Mike Robinson and David Picard (2009) also explain that photographs are a vehicle of storytelling since most tourists will talk about their holidays once back home. Whilst international visitors will be able to talk about their visit to a concentration camp memorial from the vantage point of 'seeing is believing', German visitors will relate to such a visit on a deeply personal level that one might not want to discuss. Moreover, as academic research has shown (e.g. Mitscherlich and Mitscherlich 1998), there is a reluctance to discuss the Nazi past in German families; thus, a reminder of a visit to a memorial site of the Nazi past might not be desired.

Bautzen II, however, differed significantly as visitors frequently took photographs, regardless of the cultural background. The view from the ground floor looking through the typical prison wire to the upper floors often resulted in a pause for photographs. German visitors at Bautzen II did not display the reluctance they have shown at the concentration camp memorials. Since Bautzen II was generally known in East and West Germany as the most notorious Stasi prison, taking photographs might function as a form of evidence, a recognition, that the place exists. In addition, the GDR past is not characterised by a large-scale genocide, which might make a conversation about a visit to a Stasi prison a less 'uncomfortable' one.

Debbie Lisle (2016) argues that research into tourism at sites of atrocities needs to consider the ethical and political dimensions that tourism to these sites inherently involves. We thus need to move beyond the scientific labelling, which 'dark tourism' intended to do, and reflect on the complex performance of the tourist herself. In the previous paragraphs, I have also shown that an analysis of the discourse and the narrative in memory studies lost sight of the larger significance (Sumartojo 2020). It is, after all, the visitor who engages with the site. By focusing, for instance, on the experiential nature of the visit, we observe how memories emerge, which in turn limits a state's ability to design official memory sites. Whilst the cultural framework certainly plays a significant role in how the memorial sites are viewed, the *Erfahrungshorizont* (how I remember and experience the past, Koselleck 1979) influences how a visitor negotiates the sites.

5.8 Bibliography

Arp, Agnes. 2021. 'Grauzone – Der Wert unserer Erinnerung'. Accessed 28 March 2021. www.mdr.de/kultur/podcast/feature/audio-feature-grauzone-judith-burger-102.html.

Assmann, Aleida. 2011. *Erinnerungsräume: Formen und Wandlungen des kulturellen Gedächtnisses.* 5th ed. München: C.H. Beck.

Berdahl, Daphne. 1999. *Where the World Ended: Re-Unification and Identity in the German Borderland.* Berkeley, CA: University of California Press.

Bertelsmann Stiftung, Jana Faus, Matthias Hartl, and Kai Unzicker. 2020. *30 Jahre Deutsche Einheit. Gesellschaftlicher Zusammenhalt im vereinten Deutschland.* Gütersloh: Bertelsmann Stiftung. www.bertelsmann-stiftung.de/en/publications/publication/did/30-jahre-deutsche-einheit-all.

Bilewicz, Michał, and Adrian Dominic Wojcik. 2018. 'Visiting Auschwitz: Evidence of Secondary Traumatization among High School Students'. *American Journal of Orthopsychiatry* 88 (3): 328–334.

Bird, Geoffrey R., Hilary Leighton, and Ann-Kathrin McLean. 2020. 'A Matter of Life and Death: Tourism as a Sensual Experience'. In *Tourism and Embodiment*, edited by Catherine Palmer and Hazel Andrews, 116–131. London; New York: Routledge.

Böhme, Gernot. 2017. *Atmospheric Architectures: The Aesthetics of Felt Spaces.* Translated by A.-C. Engels-Schwarzpaul. New York: Bloomsbury.

Boorstin, Daniel J. 1972. *The Image: A Guide to Pseudo-Events in America.* New York: Atheneum.

Bourdieu, Pierre. 1990. *Photography: A Middle-Brow Art.* London: Polity Press.

Clarke, David. 2019. 'Tourists As Post-Witnesses in Documentary Film: Sergei Loznitsa's *Austerlitz* (2016) and Rex Bloomstein's *KZ* (2006)'. SSRN Scholarly Paper ID 3351085. Rochester, NY: Social Science Research Network. https://papers.ssrn.com/abstract=3351085.

Cole, Tim. 1999. *Selling the Holocaust: From Auschwitz to Schindler: How History Is Bought, Packaged, and Sold.* New York: Psychology Press.

Crețan, Remus, Duncan Light, Steven Richards, and Andreea-Mihaela Dunca. 2019. 'Encountering the Victims of Romanian Communism: Young People and Empathy in a Memorial Museum'. *Eurasian Geography and Economics* 59 (5–6): 632–656.

Drozdzewski, Danielle, Sarah De Nardi, and Emma Waterton. 2016. 'Geographies of Memory, Place and Identity: Intersections in Remembering War and Conflict'. *Geography Compass* 10 (11): 447–456.

Eckmann, Monique. 2010. 'Identitäten, Zugehörigkeiten, Erinnerungsgemeinschaften. Der Dialog zwischen Vermittlungsposition und Empfangsposition'. In *Verunsichernde Orte. Selbstverständnis und Weiterbildung in der Gedenkstättenpädagogik*, edited by Barbara Thimm, Gottfried Kößler, and Susanne Ulrich, 64–69. Frankfurt am Main: Brandes & Apsel.

Eschebach, Insa. 2020. 'Die Mahn und Gedenkstätte Ravensbrück. Planungs- und baugeschichtliche Entwicklungen'. In *Gestaltete Erinnerung: 25 Jahre Bauen in der Stiftung Brandenburgische Gedenkstätten 1993–2018. Eine Dokumentation*, edited by Günter Morsch and Horst Seferens, 261–283. Berlin: Metropol Verlag.

Faulenbach, Bernd. 2019. 'Eine neue Erinnerungskultur? – Entwicklungslinien und Probleme der Gedenkstätten seit der Epochenwende 1989/90'. *Sachsenhausen Lectures* 3 (1): 1–44.

Frevert, Ute. 2013. *Emotions in History: Lost and Found.* Budapest: Central European University Press.

Gensburger, Sarah, and Sandrine Lefranc. 2020. *Beyond Memory: Can We Really Learn from the Past?* Palgrave Macmillan Memory Studies. Cham, Switzerland: Palgrave Pivot.

Giddens, Anthony. 1991. *Modernity and Self-Identity: Self and Society in the Late Modern Age.* Cambridge: Polity Press.

Groschek, Iris. 2020. 'KZ-Gedenkstätten und Social Media'. In *Kultur in Interaktion. Co-Creation im Kultursektor*, edited by Christian Holst, 105–118. Wiesbaden: Springer Gabler.

Jureit, Ulrike, and Christian Schneider. 2010. 'Unbehagen mit der Erinnerung'. In *gefühlte Opfer. Illusionen der Vergangenheitsbewältigung*, 7–16. Stuttgart: Klett-Cotta.

Kahlcke, Inga. 2017. 'Wie mit Bildern Geschichte gemacht wird: Visuelle Darstellungen des Nationalsozialismus im Geschichtsschulbuch der DDR'. *Journal of Educational Media, Memory, and Society* 9 (2): 86–109.

Kessler, Liz. 2021. 'Holocaust Stories Must Be Told, But Their Popularity Is Deeply Uncomfortable'. *The Guardian*, 27 January. www.theguardian.com/books/2021/jan/27/holocaust-stories-auschwitz-birkenau-fashion-uncomfortable.

Knigge, Volkhard. 2004. 'Museum oder Schädelstätte? Gedenkstätten als multiple Institutionen'. In *Museumsfragen. Gedenkstätten und Besucherforschung*, edited by Stiftung Haus der Geschichte der Bundesrepublik Deutschland, 17–33. Bonn: Stiftung Haus der Geschichte der Bundesrepublik Deutschland.

Koselleck, Reinhart. 1979. *Vergangene Zukunft: Zur Semantik geschichtlicher Zeiten.* Frankfurt am Main: Suhrkamp.

Kowalczuk, Ilko-Sascha. 2021. 'Das Nicht-vertrauen-können belastet Ostdeutschland bis heute'. *Berliner Zeitung*, 1 February. www.berliner-zeitung.de/zeitenwende/das-nicht-vertrauen-koennen-belastet-ostdeutschland-bis-heute-li.135563.

Kuchler, Christian. 2021. *Lernort Auschwitz: Geschichte und Rezeption schulischer Gedenkstättenfahrten 1980–2019.* Göttingen: Wallstein Verlag.

LaCapra, Dominick. 1994. *Representing the Holocaust: History, Theory, Trauma.* Ithaca: Cornell University Press.

Lennon, John, and Malcolm Foley. 2010. *Dark Tourism: The Attraction of Death and Disaster.* London; New York: Cengage Learning.

Leo, Annette. 2000a. 'Das Bild der DDR und des realen Sozialismus'. In *Zweierlei Geschichte. Lebensgeschichte und Geschichtsbewußtsein von Arbeitnehmern in West- und Ostdeutschland*, edited by Bernd Faulenbach, Annette Leo, and Klaus Weberskirch, 260–299. Essen: Klartext-Verlag.

———. 2000b. 'Das Problem der Nationalsozialistischen Vergangenheit'. In *Zweierlei Geschichte. Lebensgeschichte und Geschichtsbewußtsein von Arbeitnehmern in West- und Ostdeutschland*, edited by Bernd Faulenbach, Annette Leo, and Klaus Weberskirch, 300–347. Essen: Klartext-Verlag.

Lisle, Debbie. 2016. *Holidays in the Danger Zone.* Minneapolis: Minnesota University Press.

Mitscherlich, Alexander, and Margerete Mitscherlich. 1998. *Die Unfähigkeit zu trauern: Grundlagen kollektiven Verhaltens.* 2nd ed. München: Piper Verlag GmbH.

Muzaini, Hamzah. 2015. 'On the Matter of Forgetting and "Memory Returns"'. *Transactions of the Institute of British Geographers* 40 (1): 102–112.

Özyürek, Esra. 2019. 'Muslim Minorities as Germany's Past Future: Islam Critics, Holocaust Memory, and Immigrant Integration'. *Memory Studies*: 1–16.

Palmer, Catherine. 2017. *Being and Dwelling through Tourism: An Anthropological Perspective.* Abingdon, New York: Routledge.

Palmer, Catherine, and Hazel Andrews. 2020. 'Tourism and Embodiment: Animating the Field'. In *Tourism and Embodiment*, edited by Catherine Palmer and Hazel Andrews, 1–8. London; New York: Routledge.

Paulus, Frieder M., Laura Müller-Pinzler, Stefan Westermann, and Sören Krach. 2013. 'On the Distinction of Empathic and Vicarious Emotions'. *Frontiers in Human Neuroscience* 7 (May). https://doi.org/10.3389/fnhum.2013.00196.

Porsdorf, Friedrich. 2019. Personal communication. *Re-design of the Exhibition at Ravensbrück in 1980*, 1 July 2019.

Robinson, Mike, and David Picard. 2009. 'Moments, Magic and Memories: Photographing Tourists, Tourist Photographs and Making the World'. In *The Framed World: Tourism, Tourists and Photography*, edited by David Picard and Mike Robinson, 1–38. Farnham, England; Burlington, VT: Routledge.

Rosa, Hartmut. 2019. *Resonance: A Sociology of Our Relationship to the World.* Translated by James Wagner. Medford, MA: Polity Press.

Sattler, Victor. 2020. 'Virtuelles Holocaust-Gedenken: Remix der Erinnerung'. *Frankfurter Allgemeine Zeitung*, 16 November. www.faz.net/1.7053582.

Sierp, Aline. 2020. 'EU Memory Politics and Europe's Forgotten Colonial Past'. *Interventions* 22 (6): 686–702.

Stone, Philip R. 2018. 'Dark Tourism in an Age of Spectacular Death'. In *The Palgrave Handbook of Dark Tourism Studies*, edited by Rudi Hartmann, Richard Sharpley, Tony Seaton, Leanne White, and Philip R. Stone, 189–210. London: Palgrave Macmillan.

———. 2019. 'Dominions of Dark Tourism and "Ghosts" of the Significant Other Dead'. *Atlas Tourism and Leisure Review* 1: 5–6.

———. 2020. 'Dark Tourism and "Spectacular Death": Towards a Conceptual Framework'. *Annals of Tourism Research* 83: 2–4.

Sumartojo, Shanti. 2016. 'Commemorative Atmospheres: Memorial Sites, Collective Events and the Experience of National Identity'. *Transactions of the Institute of British Geographers* 41 (4): 541–453.

———. 2020. 'Lieux de Mémoire through the Senses: Memory, State-Sponsored History, and Sensory Experience'. In *The Routledge Handbook of Memory and Place*, edited by Sarah De Nardi, Hilary Orange, Steven High, and Eerika Koskinen-Koivisto, 249–253. London; New York: Routledge.

Sumartojo, Shanti, and Sarah Pink. 2018. *Atmospheres and the Experiential World: Theory and Methods.* London; New York: Routledge.

Tajfal, Henry, and John. C. Turner. 2004. 'The Social Identity Theory of Intergroup Behavior'. In *Political Psychology: Key Readings*, edited by John T. Jost and Jim Sidanius, 276–293. New York, Hove: Psychology Press.

Turner, Jennifer, and Kimberley Peters. 2015. 'Unlocking Carceral Atmospheres: Designing Visual/Material Encounters at the Prison Museum'. *Visual Communication* 14 (3): 309–330.

Turner, Louis, and John Ash. 1976. *The Golden Hordes: International Tourism and the Pleasure Periphery.* New York: St. Martin's Press.

Wagner, Jens-Christian. 2017. 'Gedenkstättenarbeit in Deutschland seit 1945: Eine Erfolgsgeschichte?'. *Netzwerk Erinnerung und Zukunft Hannover e.V.* (September): 5–10.

Weissman, Gary. 2004. *Fantasies of Witnessing: Postwar Efforts to Experience the Holocaust.* Ithaca: Cornell University Press.

Welzer, Harald, Sabine Moller, and Karoline Tschuggnall. 2002. *'Opa war kein Nazi':- Nationalsozialismus und Holocaust im Familiengedächtnis.* Frankfurt am Main: Fischer.

6 The future of memory in Germany

6.1 Tourism and memory

The visitor research outlined in this book has revealed the unique quality of the individual tourist experience at the different sites. Crucially, it has shown that visitors are not passive onlookers. In fact, as Kerry Whigham (2020) highlights, memorial sites intend to bring people together in a shared public space that centres around the commemoration of past atrocities. As such, visitors enact memory and actively engage with the site to create their own meanings. Dark tourism scholarship, established in the United Kingdom, has neglected the performativity of the visitor experience at memorial sites and the role of memorial sites in post-conflict societies. In the former communist Eastern Europe, for instance, visiting a memorial site is not just a postmodern experience of consumption; for visitors, the confrontation with the committed atrocities is very 'real', as the research at Bautzen II has shown (Van der Laarse 2013). In addition, visitors to memorial sites are often culturally biased and frame the memorial site according to their preconceptions (*ibid*). Within the German context, a visit to a memorial site which commemorates the Nazi past can result in conflicting emotions: for example, the shame about the atrocities committed in the name of the German nation, or the desire to come to terms with one's own family involvement in the Nazi regime, yet also rejecting any personal responsibility for Germany's past. Dark tourism's assertion that tourists relish death and the macabre does a grave injustice to the complexity of visits to such sites and to the consciousness of their visitors. In fact, the research has shown the importance of the 'social frame' (Halbwachs 1925). Visitors view memorial sites through a cultural lens: for example, the American visitors focused on the memorial plaque for the American infantry division, while Swedish visitors enjoyed the beauty of the landscape.

In memory studies, visitors to memorial sites are frequently referred to as secondary witnesses. Sara Jones (2014), for instance, argues in her research on Hohenschönhausen that the aura of the site turns the visitor into a participant in witnessing. According to Silke Arnold-de Simine (2013), it is the use of experiential architecture in combination with visual design features that produces the secondary witness. Dominick LaCapra (1994) contends that

DOI: 10.4324/9781003126836-6

'secondary witnessing' is a form of empathy that does not lead to full identification with the victim. Alison Landsberg (2009) goes even further and argues that we can acquire memories that we have not actually lived through. Moreover, the viewing of traumatic films, for example, *Schindler's List*, can encourage us to think critically about contemporary political challenges, thus building the foundation for ethical citizenship. Arnold-de Simine (2013, 42) suggests that these theories barely discuss the 'elephant in the room', which is empathy. Secondary witnessing is underpinned by the belief that an encounter with trauma is either an essential requirement or an outcome. Yet these assumptions are rarely critically examined.

My ethnographic research was the method that uncovered most of the complexities of emotions, including the fragility of empathy. At Flossenbürg, the graphic details of the death marches and the liberation led to a reaction in the form of disgust. At the House of the Wannsee Conference, visitors responded emotionally to images of Jews digging their own graves or to scenes of mass murder. The 'original' sites of suffering – such as the ramp, the shower rooms and the crematorium in Flossenbürg, or the execution path and the crematorium in Ravensbrück – were the locations at which visitors appeared to be most affected. By, for instance, being hesitant to walk into a crematorium or avoiding it altogether, visitors imagine themselves in the position of the victim, a process that Landsberg (2004) described as 'prosthetic memory'. However, 'it is the power of the direct contact with the physicality' (Popescu 2016, 276) that engenders such reactions. These emotions are also projected onto the surrounding landscape where 'buildings looked sad' or caused sleepless nights like the youth hostel in Ravensbrück. Indeed, one visitor in Ravensbrück suggested that 'these sites have a dark atmosphere that always hangs over them'. It is, therefore, the symbolic value of the memorial site that evokes such intense emotions. Hence, I agree with Daniel Reynolds (2018, 23), who argues that 'embodiment is central to the experience of Holocaust tourism'.

Phrases like 'you have to distance yourself from it, otherwise you can't cope' were expressed by some German visitors at both concentration camp memorials. This is a form of emotional distancing to avoid being overwhelmed by the physical evidence of mass murder. I therefore question Marianne Hirsch's (2008) theory of 'postmemory' in which she argues that future generations will be able to work through the trauma of the Holocaust. By creating an emotional distance, which some visitors clearly do, a barrier between the visitor and the victim is erected in order to avoid emotional exhaustion. As Aleida Assmann and Juliane Brauer (2011) emphasised, it is important to ask the question of how and why emotions are experienced at a memorial site, instead of assuming that all future generations will automatically identify with the trauma of the Holocaust.

In particular, the House of the Wannsee Conference split the audience in terms of emotional reactions. Whilst some visitors left the site in tears and described shivers running down their spines, others thought that they had seen these images before and that the exhibitions were repetitive. This was also a sentiment expressed at Ravensbrück. Thus, Hirsch's (2001) theory of photographs bridging the gap between the past and subsequent generations is not entirely correct. Indeed, Barbie Zelizer's (1998) and Susan Sontag's (2013) argument that the circulation of photographs of violence causes a banalisation of the Holocaust was certainly evident in this research. It points to a form of numbness as the more visitors experience memorial sites and exhibitions about the Nazi past, the less impact they make. Visitors who travel to multiple memorial sites develop an understanding of the 'Holocaust canon', and consequently, empathy, often the desired outcome of a memorial site, is fragile.

Emotions and the experience of trauma have so far predominantly been researched within the context of the Holocaust. How visitors encounter memorial sites that commemorate the much more recent atrocities of the communist past are rather less well evidenced. Bautzen II has shown that visitors reacted intensely to the site and used very strong expressions to describe it ('inhumane', 'degrading'). As I witnessed during the observations, at times some visitors left the site quickly. Thus, emotions at memorial sites in post-conflict societies can run high. However, Bautzen II has also shown that emotions depend significantly on the visitor's background. For instance, having experienced the GDR does not necessarily lead to empathy. The complete opposite can be the case where visitors felt that their own suffering had been worse compared to the prisoners' pain. Thus, while analysing the narratives at memorial sites in memory studies has given us a great insight into the way societies display the past, it is insufficient for an understanding of the dynamics at memorial sites.

Fritz Breithaupt (2019) explains that empathy can be blocked using a variety of methods. Familiarity with a situation prevents curiosity, which in turn can lead to a lack of empathy, for example, the visitors who claimed that they had visited several concentration camp memorials already and so did not feel a sense of empathy. In-group bias can influence empathetic reactions since we are more inclined to feel empathy with people with whom we share similarities. Sarah Gensburger's (2019) research at a Holocaust exhibition in Paris showed that visitors commented on the lack of visitors from an immigrant background, which, according to the visitors, is somewhat understandable since it is not *their* history. Prior to the interview, however, these visitors emphasised the importance of learning the lessons. Jochen Fuchs (2019), an academic at the FH (University of Applied Sciences) Magdeburg/Stendal who conducted a comparative study of students who travelled to Auschwitz and those who did not, concluded that there was only a marginal difference in students' attitudes when asked about solidarity with minority groups. In other words, the visit to Auschwitz did not result in significantly altered behaviour; hence, visits to memorial sites are not a 'miracle weapon' to combat current societal challenges. Moreover, empathy requires emotional maturity and openness, and there are a range of possibilities to explain why it might not occur. Indeed, as Crețan et al. (2019) emphasise, feelings of empathy arise from the interaction between the visitor and the physical space. Hence, a memorial site can plan for an empathetic response but cannot predetermine the outcome. In essence, the notion of a 'universal' empathetic experience at a memorial site which is prevalent in memory studies cannot be upheld during empirical research.

This research has also demonstrated that the categories of cultural memory and communicative memory are not adequate categories to fully understand the dynamics at memorial sites. Gerald Sebald (2011) argues that the transition between these categories is rarely questioned or empirically researched. Although written and visual presentations of the past are a significant part of public life since the 20th century, their role in family narratives is mostly overlooked. Recalling the exhibition design in the GDR and passing it on to the next generation illustrates that the boundaries between communicative and cultural memory are porous. In relation to the Nazi past, the *Bildgedächtnis* (memory of images) is at times so strong that it can overshadow any other narratives.

'I watched *The Boy in the Striped Pyjamas* and this is how I imagine this place to be.' This comment was made by a teenager on arrival at Ravensbrück with her parents. She later asked her mother if 'you could see those glass cases here'. Ruth Klüger (2001, 74) notably said about journeys to Auschwitz that 'he who thinks something could be found there, has brought it with him in his luggage'. Indeed, most visitors have mental images in their head and increasingly this image is Auschwitz. Visitors referred often to Auschwitz as a potential next destination for a visit to a concentration camp memorial, and those who had already visited concluded that 'Auschwitz was much worse'. The universalisation of Holocaust memory is, according to Jens-Christian Wagner (2017), the source of growing unease amongst managers in Germany, and rightly so. The consequence of this dominant narrative is that the Nazi past is reduced to the Holocaust and within the Holocaust to Auschwitz only. It is evident also in these comments: 'I did not know that there were other victim groups other than the Jews', or 'I have already been to Auschwitz and I hope the living conditions were better here [Ravensbrück]'. Michael Jeismann pessimistically concluded in 2001 that we cannot come back from the overarching narrative of Auschwitz, it is too late. The results of my research fully bear this out.

The advent of the film *Schindler's List* changed the course of Holocaust memory for good. Steven Spielberg, the film's director, pointed out that it is important 'to feel the Holocaust' (Jeismann 2001, 7). Thus, *understanding* the Holocaust is no longer enough in a postmodern society: one has to 'experience' it. Jeffrey Alexander (2002) called it the 'trauma drama'. It is not the underlying pain that causes emotional responses, it is emotionally charged media representations that engender the reactions. While the 'trauma drama' of the Holocaust also lends itself to create an emotional film, forced slave labour in the quarry at Flossenbürg is much less 'exciting'. When the visitor encounters the rather clinical and rational approach to memorialisation in Germany, s/he is disappointed. In essence, the approach to Holocaust Education that favours an emotionalisation of history by focusing on testimony (Knoch 2020) comes into conflict with Germany's *Gedenkstättenpädagogik* that champions reflective learning with an emphasis on democratic values at the memorial sites.

This research has, however, shown that the wider approach to education has led us into a cul-de-sac. For instance, the findings of a study on Holocaust education in the United Kingdom concluded that most pupils read the book *The Boy in the Striped Pyjamas* with little analysis of its fictional content (Foster et al. 2016). The study also found that knowledge about the Holocaust is rapidly in decline despite record sums being spent on education. Nevertheless, their study falls into the same trap of the Holocaust narrative: it does not analyse the knowledge about the wider Nazi past. This situation is not significantly different in Germany with one visitor explaining to me at Ravensbrück that in school her daughter only learned about the 'big' camps.

Natalie Bormann (2018), a lecturer in political science in the United States, reflected on some five-day student trips to Auschwitz, Dachau and Ravensbrück using autoethnography. She argues that students, including herself, have little knowledge about the complexity of the past when they start their journey: for example, one student commented, 'I had no idea that there was a women's camp'. Ravensbrück challenged the standard victim/perpetrator narratives, but students were also surprised by the lack of a sense of arrival at the memorial, that is, the long walk through peaceful and quiet woodlands before one reaches an equally quiet memorial site without groups of visitors in front of an *Arbeit macht frei* gate. In fact, Bormann (2018, 54) suggests that the site has a 'ghostly' appearance as it is predominantly marked by absence. Bormann's account highlights how influential the one-sided focus on Auschwitz and the Holocaust is in educational settings. It was not until the students arrived in Ravensbrück that a wider discussion about the Nazi past started to take place.

Jeismann (2001) even argues that in the United States, the memory of the Holocaust is 'out of control'. There is no better example to demonstrate this than Trump's assertion that 'fences can be a beautiful sight' (Nowthisnews 2018). This comment was met on social media with images of the Nazi era picturing barbed wire fences. It was not clear from the social media posts where the images had originated. The Nazi past, specifically the Holocaust, has become the moral highpoint from which to judge and justify current political actions and is consequently decontextualised. Trump's comment, undoubtedly inappropriate, was made in reference to the (perceived) large-scale immigration from Latin America which, after all, is not a genocide nor did he propose such measures.

The posting of concentration camp images points to another issue: the reduction of the Nazi past to a set of icons. When visitors ask at Flossenbürg what is the difference between cremation and gasification, or argue at Ravensbrück that the most interesting part of visiting a concentration camp is seeing the gas chamber, they are 'on the hunt' for 'reality'. Those icons represent the Nazi past. If they are absent, disappointment ensues. Undoubtedly, Holocaust popular culture has influenced this development, yet the memorial sites have contributed to this. In Germany, the focus of managing memorial sites is on securing traces, for instance the recovery of the foundations of the barracks in Ravensbrück, or the re-instatement of the *Appellplatz* in Flossenbürg. In addition, the story of the Nazi past is frequently explained only by commonly known objects. In fact, Wagner (2017) rather flippantly suggests that curators at memorial sites cling onto these objects for dear life. Yet these objects cannot explain the persecution in the camps and, most importantly, they inadvertently support the very fetishisation that the memorials are trying to fight against.

Memory studies scholars often concentrated on Germany's *Vergangenheitsbewältigung* by analysing public discourses such as the controversies surrounding the Holocaust memorial in Berlin, the *Wehrmachtausstellung* and the development of the memorial sites. In particular, Aleida Assmann (2011) refers to Germany's extensive memorialisation programme after years of forgetting. The individual barely features in these analyses, which is also the most significant criticism of cultural memory theories. René Lehmann's (2011) research with an East German family, a mother born in 1944 in Silesia and a daughter born in 1960 in the GDR, reveals how individuals not only are certainly aware of national narratives in relation to the GDR but also unconsciously refer to antifascist narratives they learned in the GDR. Both explain in the interview that the lack of adequate social support causes the neo-fascist tendencies in contemporary East Germany. It derives from Dimittroff's fascist totalitarian theory which argues that capitalism produces and reproduces fascist tendencies (*ibid*). Hence, state narratives are interwoven with personal experience to make sense of the present, and as such it is essential not to lose sight of the fact that it is, after all, individuals who remember the past. Consequently, research into memory must include the meaning-making processes.

An East German visitor remarked at Flossenbürg: 'Imagine, in Buchenwald they removed Ernst Thälmann and they only brought him back after protests.' This comment is worth unpacking. Firstly, there is the use of the word 'they'. This indicates the distinction between 'them' and 'us', with 'them' being the experts. It also hints at Germany's top-down approach to memorialisation which, as in this case, has alienated a large section of German society: the East Germans. The redesign of the exhibitions was informed by the results of the Enquete Commissions. During this process, the assumption was made that the change of the exhibitions would automatically lead to an altered attitude of the East Germans. Yet this approach ignored the fact that the second generation of East Germans, born and raised in the GDR, were educated according to GDR school curricula which emphasised the antifascist struggle. This generation, unlike the first and third generations, also had no comparison; they were socialised in the GDR. For them, the Westernised narrative which Germany pursued after reunification is one they do not recognise and consequently do not share. Although both Ravensbrück and Flossenbürg have exhibits on memory politics, Flossenbürg's exhibition is academically challenging, and Ravensbrück's former GDR exhibition is for most visitors just an 'art installation' (as most memorial rooms consist of sculptures). My research has highlighted that changing state narratives is not a guarantee that people will change too. In fact, if people are not included in the change process, it is a recipe for estrangement.

Secondly, there is the importance of 'roots' and 'personal identity'. When visitors travel to memorial sites, they seek a validation of the values and narratives they have grown up with. In the case of the East Germans that might be Ernst Thälmann, who was an iconic figure in the GDR. Moreover, they will refer to the Buchenwald lampshades and not the *Holocaust* TV series. Whilst Germany represents a microcosm for these 'memory wars', they happen on a much grander scale on a European level. There is a division between East and West Europe with regard to Holocaust memory. Whilst the western part of Europe had decades of getting used to the importance of Holocaust memory, the former Eastern part had not. Thus, European declarations such as the Stockholm initiative from 2000, which is based on Western values, will not solve these memory wars. In fact, in times of political crisis oppressed groups will attempt to reclaim 'narratives', visible in Europe at the moment: for example, in Hungary the 'double genocide' (communist and Nazi victims) is the preferred narrative as an attempt against the 'cultural colonisation' by the West (Van der Laarse 2013).

Personal identity also played a significant role when German visitors recalled their own memories and/or their families' involvement in the Nazi or GDR past. During memorial visits, they negotiate between the national and family narrative. This can be an emotional experience with either angry outbreaks ('I am annoyed that Germany is still held responsible') or conflicting feelings ('my loving grandmother who also supported the Nazi regime'). Identity politics, however, do not only affect German visitors. American visitors at Flossenbürg sought out the memorial plaque for the American army or related their experiences to current combat missions. For Jewish visitors, the House of the Wannsee Conference is part of an identity that signifies not only loss but also survival, whilst for German visitors it represents the most shameful part of Germany's past. Thus, identity formation plays a substantial role in memorial site visits.

Alongside the need to cement one's roots is a desire to experience the past. 'The site resembles a parkland' or 'the barracks were missing' were frequent answers given to the question 'Does the site represent the true atmosphere of a former concentration camp?' At the House of the Wannsee Conference visitors described the site as chilling and oppressive. By contrast, in Bautzen II most visitors concluded that one could not exhibit what had happened. These visitor comments reveal the complex interaction between the visitor and the memorial site, and crucially the significance of 'aura'. While the material authenticity of the site is an important factor, its setting within the wider memorial landscape is what creates meaning for the visitor. Hence, at the concentration camp memorials, the paradoxical relationship between traces of murder and picturesque landscapes *destroys* the aura. Consequently, the site feels 'inauthentic' for the visitor. In addition, comments such as 'the new visitor centre does not fit into the landscape' or 'one should not clean these places up' emphasise that the visitor expects a landscape frozen in time to an extent where they become sacred spaces. The landscape is not allowed to move on from the atrocities and nor should there be any signs of life. The findings of the research stand in stark contrast to the objectives of the German memorial sites whose focus is to secure and document the traces of the atrocities committed. Indeed, memorial sites are only supported by the national memorial scheme if they represent the historically authentic site. This is subsequently reflected in the overall management of the sites. For instance, Philipp Oswalt and Stephanie Oswalt (2000), the landscape architects at Ravensbrück, explained that the original brief for designing the memorial site explicitly stated that no buildings should be reconstructed. There was no acknowledgement in the design brief that it will eventually be the visitor who interacts with the space and uses it to construct meaning.

At Bautzen II, for most visitors, an 'authentic' experience was not important. This result is hardly surprising. As many visitors had a lived experience of the GDR, there was no need and no desire to 'experience' authenticity. Nevertheless, Charles Maier's (2002) suggestion that Communism is on the verge of becoming a 'cold' memory, whilst National Socialism seems to continuously radiate, was premature. National Socialism is certainly a memory that dominates the overarching narratives, yet GDR memory, even in 2021, is by no means 'settled'. It remains an emotional topic within contemporary Germany. This was particularly evident in the 2018 dismissal of Dr Hubertus Knabe as director of Hohenschönhausen Stasi prison memorial. This sparked a heated debate about the failure of GDR memorialisation with Ilko-Sascha Kowalczuk (2018) concluding that 'managers at the [GDR] memorial sites are only there because of their biography, not because they are qualified'. Saxony's integration and equalities minister, Petra Köpping (2018) published a book entitled *Integriert doch erst einmal uns – eine Streitschift für den Osten* (Integrate us first – a polemic pamphlet for East Germany). She argues that, when dealing with GDR memory, one needs to consider the economic crisis in East Germany and the impact it had on people's personal lives in the 1990s.

My Bautzen research results have shown that the legacy of the GDR is very much a part of people's lives. The visit sparked memories of Stasi oppression, restricted life opportunities and 'grey' facades. Visitors also reflected on what 'could have been' if official requests to leave the GDR had gone wrong. The conversation I had with a former employee of the West German consulate in the GDR about West German prisoners at Bautzen very quickly attracted a number of visitors who were eager to hear his story. Equally, an event organised by the memorial team on the topic 'The GDR prison system', led by the former manager at Bautzen I who openly talked about his work as a Stasi informant, attracted a large audience. At times, visitors commented that they had wrestled with themselves over whether they should come to Bautzen at all. It is perhaps the increasing time distance that now allows people to reflect on their own memories of life in the GDR. There were, of course, also visitors who were conflicted: they recognised the human rights abuses, yet they themselves led 'normal' lives. For some visitors, Bautzen II was merely a site where one could gaze on the 'evil' Stasi or could finally see a prison. This behaviour, together with comments from West German visitors recalling great summer holidays with relatives in the GDR, highlights that coming to terms with the GDR past is by no means merely an East German problem. There is a tendency in contemporary Germany to 'eradicate' positive memories of the GDR and to focus on the SED dictatorship with specific attention to the East Germans. Yet a West German visitor who proudly announces that he has already sat at Mielke's desk obviously is not someone who understood the GDR past. In addition, as Mary Fulbrook had already pointed out in 2007, the simplification of GDR history to Stasi victims and the Berlin Wall does not contribute to a critical analysis of the GDR past (Hesselmann 2007). Moreover, as Habbo Knoch (2020) highlights, there is no consensus on how to exhibit the GDR past at memorial sites, also reflected in the fact that there are numerous publications that discuss *Gedenkstättenpädagogik* in relation to the Nazi past but barely any that address the GDR. Considering my research results, it is essential to take a closer look at the educational strategies at the memorial sites that commemorate the GDR.

Reviewing the research results, I suggest that 'memory tourism' is a term which more accurately explains this form of tourism. The term also detracts from the negative portrayal of the tourist to memorial sites as voyeuristic or shallow. Of course, there were visitors whose behaviour could be judged as superficial, especially when they only chase Holocaust icons, or refer to 'Alcatraz' in Bautzen. But these tourists are also, to an extent, engaging with memories. Indeed, these examples emphasise the blurring of boundaries between fact and fiction which Alison Landsberg (2004) notes in her concept of prosthetic memory. Both the Holocaust and Alcatraz are featured in a variety of emotionally charged cinematic experiences, thus the visitor is familiar with these 'memories'.

Creating a 'typology' for the average visitor is challenging or even, as I argue, unhelpful. In fact, as Judith Mastai (2007) highlights, there is no such thing as *a* visitor. She argues that traditionally visitor research in museums is conducted according to socio-economic profiling with the aim of creating products, mainly blockbuster exhibitions, to draw in visitors and increase secondary spend (*ibid*). Memorial sites, however, are not merely businesses. Indeed, in Germany they are not businesses at all. Furthermore, tourism theory analyses tourist motivation according to push and pull factors, for example, the attractiveness of a destination. In essence, dark tourism theory has followed this narrative by arguing that the 'spectacle of death' is a pull factor for rising visitor numbers at memorial sites. Western societies, however, value individualisation and self-determination, which means that a person is in control of his/her own decision-making processes. Thus, tourists are not passive consumers influenced purely by external values. Their need for self-actualisation has a significant impact on whether they visit a memorial site and, once they are there, how they 'experience' the site. Therefore, instead of distinguishing visitors according to socio-economic factors, I suggest differentiating them by extrinsic or intrinsic motivation.

Intrinsic motivation is characterised by a desire for self-actualisation, rather than by external pressures. For humans, intrinsic motivation is important as they are naturally curious and inquisitive, displaying a willingness to learn and explore (Ryan and Deci 2000). Visitors to memorial sites who are intrinsically motivated will therefore place curiosity and knowledge at the heart of the visit. For instance, a visitor who realises that s/he has learned a one-sided version of history in the GDR, and therefore decides to visit Flossenbürg, is an intrinsically motivated visitor. For these visitors, the memorial site forms an important backdrop to acknowledging Germany's challenging past and to deal with one's own and/or one's family's involvement in the Nazi past.

In contrast, extrinsic motivation is driven by external pressures and a reward system, for example, the student who studies hard because his/her parents expect a high grade. Thus, extrinsic motivation is characterised by an activity that is done to attain a certain outcome, not simply for enjoyment (*ibid*). An extrinsically motivated visitor is driven by a sense that seeing is believing. The rewards of such a visit are the barbed wire fence posts or the crematorium. These are visitors who are emotionally removed from this past and display signs of voyeurism. For instance, the visitors at Bautzen II who make comparisons to Alcatraz, or the young people at Flossenbürg who think that Auschwitz is more interesting as 'one can see the scratch marks on the wall'. This type of visitor is, however, not the majority and, with a thought-provoking exhibition, one can motivate these visitors to think beyond their preconceived perceptions. In fact, research in educational psychology has shown that if learners feel connected and possess a level of agency, then motivation can switch from the extrinsic to the intrinsic (*ibid*). However, the information-intensive approach to exhibition design in Germany prevents this 'transformation'; hence, I will address improvements to exhibition design in the second part of this chapter.

Academic and public discourse could, of course, continue attacking tourists to memorial sites, which is the easy route since it avoids confronting societal challenges. There are increasing debates about the presentation of memorial sites of the Nazi past in an era without eyewitnesses and the rising far-right tendencies. The approach to GDR memorialisation is equally contested, with Kowalczuk (2018) concluding in an article that uncomfortable questions in relation to coming to terms with the GDR need to be asked. The issue of commemorating communist atrocities also extends beyond the boundaries of Germany, with memory wars raging in the former Eastern bloc, leading to the introduction of the Annual Day of Communist Victims (23 August) as a challenge to a globalised Holocaust memory discourse. Thus, memorial sites have a much wider role to play in contemporary societies than just being tourist attractions. Indeed, in light of the rise of the far-right and antisemitism in Europe (Grindheim 2019), the question needs to be asked whether memorial sites ever can be agents of change by critically engaging visitors, and if they are not how one might achieve this.

6.2 The future of German memory culture

In an interview with the journalist Tim Adams (2016), Neil MacGregor, the former head of the British Museum, said: 'The thing I continue to find striking is that in the centre of Berlin you keep coming across monuments to national shame. I think that is unique in the world.' Germany is often praised for its memory culture that does not shy away from dealing with the atrocities committed in the name of the German nation. The British historian Timothy Garton Ash even speaks of a German DIN (German industrial norms) norm in how to confront a 'difficult' past (Faulenbach 2019). With the rise of the Black Lives Matter movement, praise for Germany's memory culture reached new heights when the American philosopher Susan Neiman (2019) published her book entitled *Learning from the Germans: Confronting Race and the Memory of Evil*, and even she admits that people in Germany laughed when she mentioned the book title.

Within Germany the established memory culture comes increasingly under pressure. In fact, if one scratches the surface, serious cracks appear. Wagner (2017) confirms that, while outwardly Germany is considered to be exemplary in terms of addressing challenging pasts, amongst the memorial managers there is a growing unease about representations of the Nazi past. Firstly, Wagner (*ibid*) argues that the focus on using eyewitnesses has diminished the ability to critically analyse historical facts and has shifted the attention towards remembrance. This leads to obscure commemorative events such as the performance of the government in Thuringia which honoured the victims of the Nazi past in the same breath as the victims of the 2004 Tsunami, as if the Holocaust was a natural disaster. Secondly, the victim focus, specifically on the Jewish victims, marginalises other victim groups. As a specific example, Wagner (*ibid*) highlights an article in the *Celle Zeitung* about French resistance fighters at Bergen-Belsen with the headline 'witness to Jewish suffering'. On a closer look, however, none of the mentioned resistance fighters were in fact Jewish. Thirdly, there is a tendency at German memorial sites to focus on objects, documents and eyewitnesses to explain the Nazi past. Yet in an age where visitors have less and less knowledge, this exhibition design does not make sense. Gerd Kühling designed a temporary exhibition under the title *Repressed Memory – how West Berlin dealt with Sites of Nazi Perpetrators* and asked visitors at the end 'What is missing in today's culture of remembrance?' Early answers indicate that visitors are in fact interested in the 'grey areas' and want to be challenged. Indeed, visitors to his exhibition, for instance, frequently mentioned marginalised victim groups such as the 'asocials' (Kühling 2017).

Dana Giesecke and Harald Welzer demanded already in 2012 that Germany's memory culture requires a renovation. Since the 1970s, Germany pursued a commemorative culture that has the victim at its heart and thus attempts to explain the Nazi regime from its endpoint. Yet the victim's perspective cannot explain how Germany's society descended into a vicious circle of discrimination, violence and hatred, and how millions of Germans stood by as their

Jewish neighbours were deported. Hence, one needs to analyse the mechanisms of violence and how it penetrates societies if one wants to instil democratic values in a younger generation. It would also involve the dismantling of the strict categories of perpetrator, victim and bystander as the boundaries are often fluid, for instance an SS man who married a Jew to avoid deportation (*ibid*). Giesecke and Welzer (*ibid*) also rightly note that the speed at which the former Yugoslavia suddenly engaged in ethnic cleansing and mass murder shows that societies are not as resilient as they might appear.

The dismissal of Dr Hubertus Knabe from his post as memorial manager of Hohenschönhausen caused an intense debate about exhibition practices at the memorial sites that commemorate the GDR. The core criticism was that the memorial site left very little room for visitors to form their own opinions, a clear violation of Germany's principles of political education. In July 2020, the managing director of Saxony's Memorial Trust, Siegfried Reiprich, compared the Covid-19 protests in Stuttgart to the antisemitic November pogroms in 1938 and later complained on his Twitter account about white people soon becoming a minority in Europe. This prompted an open letter to Saxony's Ministry for Education and Culture by Germany's overarching memorial working group demanding Reiprich's removal. He was subsequently permanently suspended at the end of July 2020. Claudia Maicher, MP for the Greens in Saxony's parliament, claimed that Reiprich had not supported the principles of Germany's memory culture for a long time and had hampered the development of Saxony's memorial foundation (MDR 2020). Considering that Bautzen II is managed by Saxony's memorial foundation and struggles with the heated political situation locally, not all is well with Germany's memory culture when the managing director did not support the core values of the foundation, yet enjoyed political support until a public outcry.

At around the same time of Siegfried Reiprich's suspension, another public debate erupted surrounding the South African-based Cameroonian postcolonial thinker Achille Mbembe, who was accused of antisemitic writings. The German antisemitism commissioner, Felix Klein, demanded the cancellation of Mbembe's invitation as a speaker at the festival *Ruhrtriannale*, but the Covid-19 pandemic luckily intervened prior to the official de-invitation. Since then, a heated debate dominates the German academic and public landscape. Arguments in the press, on social media and on blogs culminated in the founding of the network *Weltoffenheit* (cosmopolitanism), which claims that Germany's prohibition of the BDS (Boycott, Divestment, Sanctions) stifles public debate in art, culture and academia. At the heart of this intense debate is Germany's insistence on the singularity of the Holocaust; equating it with other atrocities, such as the genocide in Namibia, is out of the question, which is symptomatic of Germany's inability to confront the colonial past (Rothberg 2020). Aside from the debate about the singularity of the Holocaust, an article in *Der Spiegel* highlights that Germany's way of commemorating the Nazi past cannot continue in its current format. The author describes how a Syrian refugee was taken on a trip to a concentration camp memorial as part of her German course (Amjahid 2021). Once onsite, the Syrian woman could barely hold it together as her own traumatic memories resurfaced; the course leader did not notice her as she was proudly explaining Germany's attempts to come to terms with the Nazi past. Germany will need to address its more diverse society whose family memories do not contain a relationship to the Nazi past, and whose ancestors do not belong to the Nazi perpetrators.

Germany was given little respite after the Mbembe debate: with the 30th anniversary of German reunification on 3 October 2020, a new debate emerged surrounding the legacy of the GDR, and in particular the social and economic transformation of East Germany following the fall of the Berlin Wall. A commission consisting of 22 members coming from politics, academia, the economy and culture reflected on Germany's process of reunification. They concluded that more emphasis should be put on the enormous efforts East Germans made to cope with the political and economic transformation (Bertelsmann Stiftung et al. 2020). In addition, this experience should be harnessed to research contemporary societal challenges that will inevitably lead to transformation (e.g. climate change) by launching a research centre. Moreover, victims of the GDR regime should be better compensated, and additional psychological support should be offered. The 224-page report also makes a series of recommendations in how to combat East Germany's sharp demographic decline, which is a side effect of millions of young people leaving the area in the wake of reunification. Thus, the focus seems to shift from the GDR itself to its aftermath.

In 2019, the (West) German journalist Wolfram Ette (2019) visited the temporary exhibition *Wende* (political change in 1989) in his new hometown Chemnitz. He was fascinated by the number of exhibits that showed the multitude of publications of the 1980s GDR in which people expressed their desire for reform. In particular, a samisdat (word used to describe an underground publication) captured his interest as it described the three fundamental societal fears: fear of foreigners, fear of change and fear of critical thinking. Much to his disappointment, he could only read the first two pages as the object was behind glass and a copy for general use was not provided. Ette's description of the exhibition points to two key issues: the mistaken portrayal of passive GDR citizens who accepted uncritically the regime, and a lack of engagement with museum visitors for whom the fall of the Berlin Wall is not a personal history. The significant potential to engage tourists with GDR history beyond the Stasi and dictatorship paradigm is currently being missed.

6.3 Exhibition design

'I have seen these images elsewhere', 'it is all repetitive' and 'I never read copious amounts of texts'; these are all statements made by visitors. In particular, Ravensbrück's exhibition was experienced as 'too much'. Wannsee's approach to exhibition design caused similar reactions. When one compares the exhibitions at Flossenbürg and Ravensbrück, one cannot fail to notice the 'repetition' effect as at times the same headlines are used. The exhibitions also follow a standardised approach consisting of the following themes: construction of the camps, victim groups, living conditions and liberation. Dachau's exhibition concept is exactly the same, so that way the victim is at the heart of the exhibition (Eberle 2004). *Der Spiegel* (Thöne 2018) argued that Germany is 'stuck' in the morality of the 1968 generation. This generation was the one which was old enough to remember the war, but too young to have taken part in the Nazi crimes. Increasingly disillusioned by post-war West German politics, they saw a direct connection between Auschwitz and the crimes committed in Vietnam. Germany's culture of remembrance has significantly moved on since then; current and subsequent generations do not require the same level of enlightenment. In addition, the next generations will neither have a direct connection to the Second World War nor know someone who has experienced it.

The uniform design of exhibitions, combined with the top-down approach, will not reach most visitors in the 21st century. Ravensbrück's, Wannsee's and Flossenbürg's exhibition design is based on behaviourist theories of learning: a constructor-centred teaching strategy that focuses on the retention of information; it is a strategy often found at museums in general. In contrast, constructivist learning theory has shown that learning is an active process and new information needs to be linked to prior knowledge. A visitor is not an empty container which can be filled with knowledge. S/he needs to be given the opportunity to actively take part in the process. Generally, museums are great places to foster such a learning style as they provide the visitor with the possibilities of exploring content according to their choice. Yet at the German memorial sites, this choice has been taken away from the visitor and replaced by an information-intensive exhibition. This authoritative style leaves little room for a dialogue with the visitors, treating them as part of a uniform mass. But, as the visitor research in this project has demonstrated, the visitor arrives at the memorial site with a considerable amount of baggage, containing the cultural background, life experiences, personal upbringing and increasingly an expectation of what a concentration camp ought to look like. If memorial sites see themselves as significant players in the landscape of political education in Germany, then they will need to start evaluating the contributions the exhibitions make in enabling visitors to think critically. Moreover, if memorial sites now see themselves as *Zeithistorische Museen* (museums of contemporary history, Eschebach 2020), then a behaviourist style of exhibition design is not enough.

Although largely based on her experience of science museums, Deborah Perry (2012) developed a model of how to design exhibitions that intrinsically motivate visitors, consisting of the following components: communication, curiosity, confidence, challenge, control and play. I want to focus here on the elements of communication, curiosity, challenge and control as they are particularly relevant for visits to memorial sites. Communication is a crucial part of the museum visit, with museums often speaking in a unidirectional voice: that is, there is no dialogue with the visitor which is certainly the dominant form of communication with the visitor at the memorial sites. The exhibition *Kriegskinder* (war children) in Pirna/Saxony in 2020 demonstrated how a dialogue with the visitor can be created, even when exhibiting difficult subjects. At the beginning of the exhibition, the visitor could read the following introduction:

> Germany caused World War II. Unmeasurable pain and violence were inflicted on those persecuted, deported, murdered or forced to work. Pain is however always personal and cannot be offset with other people's suffering. It remains in our memories and influences people's attitudes and thinking. This exhibition shows the voices of the war children of Pirna and surrounding area who were silent for years but whose experiences belong to the local memory of the region (my translation of the introductory exhibition panel).

Of course, this text not only shows Germany's challenges of exhibiting German suffering but also creates immediately a connection with the visitor by using phrases such as 'our memories' or the 'region'. Later on in the exhibition, the design team highlighted that some statements could be considered controversial (e.g. the eyewitness who stated that he enjoyed the social activities the Hitler Youth provided), but the designers decided to leave these statements unaltered, allowing visitors to form their own opinions and therefore being part of the dialogue. Here, visitors are given space to reflect independently on the exhibits, and judging from my brief review of the visitor comments book this concept achieved its aims. Visitors praised the engaging, thoughtful exhibition.

Communication is, however, not only restricted to the dialogues between museum staff and visitors, it should also focus on conversations between visitors; a museum visit is more meaningful if it encourages collaboration. There were numerous incidences in all of my case studies where families wanted to engage their children with a very difficult part of German history, but the memorial sites let them down. In fact, at the House of the Wannsee Conference, one visitor complained about another visitor that she was too loud when trying to explain certain aspects of the Nazi past to her daughter. In a publication on *Gedenkstättenpädagogik* published in 2015, there are hardly any chapters that specifically address informal learning opportunities for individual visitors despite the fact that the German memorial sites see themselves as *Lernorte* (Gryglewski *et al*, 2015). Heidi Behrens-Cobet already highlighted in 1998 that, although adults are the largest groups at German memorial sites, they are a blind spot in the pedagogical offers, and Germany is mostly still in this situation today.

Alongside the need for improving communication, visitors need to be challenged beyond their preexisting knowledge if exhibitions are to have an impact. Germany's victim-centric exhibitions focus on multiperspectivity which recognises the different victim groups and their individual experiences, but neglects the grey areas. Prisoner hierarchies and conflicts between different ethnic groups do not feature strongly in the current exhibition at Ravensbrück, thus often giving the impression of mutual solidarity between prisoner groups. Rydén (2018, 518), who researches the Ravensbrück archive at Lund University/Sweden, quotes an interview with a survivor who exclaimed: 'camaraderie did not exist. The nationalities fought each other. The fight for survival was most brutal and crude.' A book published by Austrian communist victims in 1945 also emphasises the hostility between the communist victim group and the asocials, with the women claiming that the asocials did not even deserve the title 'women' (Bruha et al. 1945). Prompting visitors' thinking by highlighting such conflicts makes them think critically and moves beyond the standardised victim, perpetrator, bystander paradigm.

The new generation of visitors is also asking new questions, for instance about the connection between the camp and the local community, the low prosecution rates and why the international community did not intervene at an earlier stage. Even if, as a memorial site, it is challenging to find answers to these questions, it is still important to address them. Exhibiting the links with the local community will require substantial research into the various connections. It will also mean that harrowing stories such as the sexual violence after liberation need confronting. Representing the German as a victim is highly controversial and I am fully aware of the criticisms associated with such an approach, not to mention the potential uproar it could cause amongst victim groups. Nevertheless, I think it is a topic which has impacted local communities and deserves attention. Annette Leo and Peter Grätz (2008), for instance, highlight in their book on Ravensbrück the woman who remembers the prisoners passing by her house and becoming a victim of a sexual assault herself at the hands of the Soviet Soldiers at the end of the Second World War. Developing exhibitions that challenge the standard narratives will unsettle visitors. This will also emphasise that the victim, perpetrator and bystander narratives are too simplistic and that the boundaries between these roles can be blurred. Whilst the local community was certainly a bystander, it was also turned into a victim by the violence it experienced after liberation. Posing challenging questions such as 'Should a woman's body be part of warfare?' will create discussions without providing the answer. These open questions will 'unsettle' a visitor, and it is exactly this 'empathic unsettlement', as LaCapra (1994) called it, that we need to create in these exhibitions.

In fact, Ravensbrück's new female perpetrator exhibition, opened in 2020, is starting to implement the notion of 'unsettlement' by exhibiting 'grey areas' (Erpel and Eschebach 2020). Since Ravensbrück was a 'training centre' for female staff at concentration camps, it shows women's career developments within the concentration camp system. It challenges the perception that Nazi crimes were predominantly committed by men and reflects critically on the low prosecution rate since most women were seamlessly re-integrated into West and East German society. In addition, a section of the exhibition addresses the highly sexualised portrayal of Nazi women as blond beasts, for example, women were portrayed as sadistic beasts in the *Men's Adventure Magazines* in the United States, in the *Stalag* magazines in Israel and in Italian *SadicoNazista* films

(Sommer 2020). These pornographic representations of female Nazi perpetrators persist to this day in popular culture, which made Ravensbrück's management team want to address this issue in the exhibition; they anticipate that it will engender serious debate.

Challenging visitors in exhibitions will also become increasingly difficult with content now often freely available through the Internet (including digital archives), which also changes the role of the memorial site considerably. For instance, Maria Zalewska (2018) points out in an article about social media at Auschwitz-Birkenau that the authority memorial sites used to possess is eroded as they have no control over what and in which way content is shared on social media platforms. Thus, German memorial sites now operate in a very different society compared to 40 years ago, and this is reflected in comments such as 'I can read it all at home' or 'I never read anything in exhibitions'. The pandemic, however, has shown that the German memorial sites were not prepared for the 'digital revolution'. Whilst Auschwitz-Birkenau and Yad Vashem had a social media presence for a long time in addition to digital learning materials, the German memorial sites viewed this development with scepticism and shied away from it. When the pandemic led to the school closure in Germany, the German public TV channel WDR added the film *Auschwitz – Das ehemalige Konzentrationslager in 360°* (Auschwitz – the former concentration camp in 360°) to its learning platform, and it emerged that this film has been used for some time in German schools (Kuchler 2021). Thus, the new generation will now not only be influenced by media representations but also by virtual representations. As all of this content is now available at the click of a button, the visitor does not expect an exhibition that follows a linear, predictable narrative: construction of the camp, conditions onsite, victim groups, SS staff and liberation. Flossenbürg's and Ravensbrück's exhibitions follow this pattern, at times even using the same headlines. In many ways, the new exhibition at the House of the Wannsee Conference attempted to break with the predictable format of exhibition design by showing contemporary forms of discrimination, which was immediately thwarted by a public outcry, suggesting that one cannot compare the Holocaust with a sign warning male refugees to enter a public swimming pool alone. This outrage, however, misses the crucial point that if we want these exhibitions to have an impact then they need to be relevant for today's visitors, otherwise we will continue to see visitors saying 'the visit did not make me review my own behaviour', as I have shown in this research.

Perry (2012) explains that visitors are drawn to exhibits that stimulate their intellectual curiosity, and a particular important strategy to achieve this form of curiosity is to contradict what people already know. By exhibiting seemingly ordinary women in the new perpetrator exhibition, Ravensbrück, in essence, uses curiosity to attract visitors and to move them beyond the notion of the gentle female sex. At the *Kriegskinder* exhibition in Pirna, one exhibit showed a book with the title *Rasse und Seele: eine Einführung in die Gegenwart* (*Race and Soul: An Introduction to the Present*). This book was printed in 1926 and discussed the different racial souls, for example, the northern or the oriental soul.

This research was later used by the Nazi regime to develop the pseudoscience of the Aryan race, which had its own school subject during the Nazi regime. Most visitors would have been familiar with the concept of the Aryan race, but probably would have never encountered this book; so this exhibit would have sparked their curiosity. Sometimes, museum exhibitions can arouse curiosity, but the design itself can still leave visitors disappointed, as the example of the journalist in the *Wende* exhibition in Chemnitz emphasised. He was keen to read the samisdat but the *Vitrinenprinzip* (placing the object behind glass), as he called it, did not allow him to pursue his interest (Ette 2019). In this case, a copy of the original for general use would have significantly enhanced the meaning of this visit. Thus, the development of intrinsically motivating exhibitions does not always require the newest technology, designing exhibitions with the visitor in mind can make a difference.

In addition, visitors can be intellectually challenged by not providing answers. At Flossenbürg, the local community is portrayed as having rejected the war crimes trials and/or denied that these atrocities have ever taken place. It encouraged Jörg Skriebeleit, the manager for Flossenbürg, to include the images of the death marches (made by Czech residents) in the exhibition to demonstrate that the 'Germans looked away'. However, such an assertion is problematic. The lack of German evidence about the death marches could be due to multiple reasons, such as ignorance, fear or simply no access to a camera. The exhibitions make moral judgements about the behaviour of the local community from our present point of view. If the aim of the memorial sites, however, is political education, then what is controversial needs to remain controversial. This will mean representing different perspectives and not taking the responsibility away from the visitor (Wagner 2017).

Although all of my case studies allowed visitors to explore the sites for themselves, especially at Ravensbrück, the multitude of exhibitions, with the largest one in the *Kommandantur* consisting of 15 rooms over two floors appeared to be overwhelming for some visitors. There was a sense that they had to visit all exhibition spaces, yet exhaustion arose already after the *Kommandantur*. Moreover, Ravensbrück's exhibition concept envisages a visit to the *Kommandantur* first, as an introduction to the rest of the site, with a visit to the other exhibitions if the time permits (Eschebach 2020). However, visitors do not necessarily follow this recommended route. If they, for instance, visit only the former GDR complex, they will be largely unaware of the politicisation of the Nazi past in the GDR. Most exhibitions at memorial sites are too large to visit in the space of a few hours, which is why it is important to give visitors a sense of control. As Perry (2012) explains, most visitors will require additional information in order not to become weighed down by the number of choices they have to make and/or by the vastness of the site. Creating exhibitions that address different visitor needs calls for a multilayered approach. For instance, some visitors might be drawn to Ravensbrück's later use as a Soviet army base, whilst others are interested in the Uckermark camp for young girls. By providing different levels of information or routes, the visitor is empowered to choose the content s/he wants to engage with, which might lead to more satisfactory visits since visitors can choose the content according to their own interests. In addition, my research has shown that the landscapes are a significant part of the overall experience of visiting memorial sites, thus engaging visitors with the wider landscapes through the design of specific routes could engage those visitors for whom the atmosphere is a motivating factor.

6.4 The landscape

'The landscape is not responsible for what happened here.' This is a statement projected onto the wall at the exhibition *Was bleibt. Nachwirkungen des Konzentrationslagers Flossenbürg* (What remains. The Aftermath of the Flossenbürg Concentration Camp). It reflects the sentiments of the local population who feel that the landscape is spoilt by keeping the memory of the atrocities alive. The landscape architect who designed the first memorial, *Tal des Todes* (valley of death), at Flossenbürg suggested that the memorial needs to be embedded into the wider landscape without disturbing the beauty of the surrounding area. Nature has the unique ability to transform former spaces of suffering and can be used to either hide or expose memories. It then requires historical research to uncover such memories, and once they come to the surface, the landscape's innocence is changed for good. Suddenly, nature is in the way of preserving the memories.

Hence, in Flossenbürg and Ravensbrück trees were removed if they were perceived to be in conflict with the memorial's historical structures. On the other hand, nature was used to mark camp boundaries by planting trees. The curatorial decisions on how to design memorial landscapes barely receive any attention. A very rare research project on human interventions at memorial sites is Charlesworth and Addis' (2002) study at Auschwitz and Plaszow. They highlighted that active grassland management at Auschwitz did not take place until the Pope's visit in 1979. It was not until 1992 that an ecologist was appointed to manage the landscape at Auschwitz, and controversially a process of chemical weeding commenced. The key aim was to recreate the landscape of 1943/44. Subsequently, the meadow grasslands were declared inauthentic and destroyed, resulting in the removal of plant species that are threatened with extinction. Thus, the landscape now suffers. Considering the pressures to nature, with more than 26,000 species under threat of extinction (Watts 2018) such interventions are difficult to comprehend, and it is surprising that these actions have not caught the attention of environmentalists.

Research into memorial landscape design and the visitor response is rare (Rapson 2015). Thus, the ethnographic research as part of this project has for the first time revealed the complex relationship between the visitor and the memorial landscape. Comments such as 'every stone is a destiny' or 'the trees would have been here at the time' signify how visitors create meaning by using natural features. At Ravensbrück, the clinker surface is mostly experienced as an 'in-between' space to walk from the *Kommandantur* to the former textile factory. During my fieldwork in July, it was at times hot and the surface reflected the heat, so for most visitors it was just a strenuous and bleak walk. One could, of course, argue that it mimics the hard living conditions in the camp, but visitors barely stopped to read the different labels and at times decided not to walk across the site in the first place. Hence, Oswalt and Oswalt's (2000) assertion that one does not know at the design stage how a visitor will respond is correct.

In geographical research, it is well known that landscapes have a cultural meaning. With the ambitions to expand both Flossenbürg and Ravensbrück by including the former quarry at Flossenbürg and the 'zone of misery' (south side) at Ravensbrück, curatorial decisions about landscape design will become an essential consideration. Both areas are marked by absence. There are hardly any physical traces of the former suffering. In the case of Ravensbrück, a decision was made to avoid an aestheticisation of the landscape by retaining the derelict grounds and focusing on the 'readability of the landscape' through markers (Eschebach 2020). However, Insa Eschebach (*ibid*) speculates whether one might make a virtue out of necessity since today's financial resources are considerably lower than those available in the 1990s.

If memorial sites want visitors to have a better understanding of the spatial dimensions of the site, then interpreting the landscape is vital. Alberto Giordano and Tim Cole (2018) suggest, albeit in a different context, a geographical information system of the place. In order to portray the camp system, one needs to be able to 'walk in and out of buildings'. Hence, creating location-specific interactive maps which link the visitor to his/her current standpoint using qualitative data, such as images or the voices of survivors, will interpret absence and will allow visitors to gain a sense of place. Another method could be the creation of 'soundscapes'. At strategic locations, sounds, such as working in the factory or conversations in barracks, could provide the aura visitors seek. Using sensory methods might also engage visitors who do not want to visit exhibitions. Such an approach, however, is not without controversy: for instance what about the sound of women being shot at Ravensbrück? Or victims being thrown down the ramp at Flossenbürg? These are the ethical dimensions that need to be considered with any of these developments, and I would argue against such sounds. The victims would be subject to fetishisation and their original suffering would be 'flattened' to make it accessible to the visitor. Such portrayals would also emotionally overwhelm visitors and would not create a learning environment. However, by creating a palimpsest of a landscape, the visitor could be encouraged to slow down and understand the wider narrative. Doreen Massey (2005) has emphasised that space cannot just be read as a fixed point in time, but one must be aware of its transformation. Audio walks have the ability to create a form of 'embodied listening', forcing people to negotiate the current landscape while also engaging with the 'memoryscape' (High 2013). Thus, memory is able to 'cut through the layers' and show how time-bound places are (Klüger 2001)

Such interventions also 'return' the victims into the landscape. Memorial landscapes tend to be perpetrator-dominated as the visitor gazes at buildings that were constructed either by the SS or by forced labour (Giordano and Cole 2018). The voices of the victims are noticeably absent. In fact, the only time visitors encounter the victims is in the 'perpetrator buildings'. The daily struggle for survival was, however, characterised by movement: forced labour, *Appells*, moving between barracks. Hence, current memorialisation at Flossenbürg and Ravensbrück represents a fragmented portrayal of life in a camp. Flossenbürg's memorial landscape also excludes the quarry and the SS accommodation due to the quarry still being in operation and the SS houses being private residences. At Ravensbrück, the former Siemens area, the location of the tent (which was erected in the final months with the most awful living conditions) and the Uckermark camp are also excluded. Thus, audio walks can bring inaccessible places back into the memorial landscape.

The audio walk project at Gusen/Austria, a subcamp of Mauthausen, illustrates how artistic installations through sound and voices can create an opportunity for an emotional connection without crossing ethical boundaries. Apart from a memorial erected in the 1960s and a visitor centre opened in 2004, nothing remains of the Gusen subcamp. Much like Flossenbürg, the former barracks were removed and replaced by an Austrian housing development. The audio walk guides visitors through the village where 30 voices of survivors, residents at the time of the camp's existence, contemporary residents and SS staff interplay with specially designed soundtracks. The visitor only hears fragments of those voices; hence, it is not a complete historical analysis of the Gusen subcamp. A narrator, Traudl, guides the visitor through the contemporary landscape, forcing the walker to imagine the violent past while seeing a peaceful, Austrian village. Tanja Schult (2020) notes the technical brilliance of the audio walk as it is choreographed in such a way that the walker stands in front of the former brothel just in time when one of the voices talks about the brothel. The artist, Christoph Mayer, also refrained from offering the visitor an identification with the victims since the different voices have complex characters, including the perpetrators. It confronts the visitor with an uncomfortable past and leaves it open, thus not offering the 'feel-good factor' that this past is thankfully behind us. Traudl becomes increasingly disturbing throughout the walk as she explains her 'normal life' while living next door to a concentration camp; thus, Schult (*ibid*) argues that the audio walk has the ability to confront us with ourselves. Most of us are predominantly interested in our own well-being and we all have an enormous capacity to ignore atrocities, even if they occur in front of us. Whilst the walk is not a substitute for the historical learning, by featuring those different perspectives the visitor gets a sense of the complexity of a concentration camp and its links with the local community. And visitors are encouraged to form their own opinions rather than being instructed. It is precisely these soundscapes that provide visitors with an impression of the daily hardship of camp life which Frankl (1946) described so vividly in his autobiography *Ein Psychologe überlebt das KZ* (In Search for Meaning). Such artistic interventions could also draw attention to the complexities of Ravensbrück and Flossenbürg. Juxtaposing the voices of residents, SS guards and survivors whilst guiding the visitor through the wider landscape would create the space for visitors to critically analyse the site. For instance, at Ravensbrück, one could include the area of the former Siemens factory, providing the visitor with a much greater sense of the hardship. Equally, at Flossenbürg, incorporating the surrounding area of the SS houses and the quarry would encourage a wider perspective. However, I would not stop at the point of liberation in 1945.

At Ravensbrück, the contextualisation of the GDR's approach to memorialisation (walking from the dark to the light) is absent. Effectively, Ravensbrück consists of two memorials now, the GDR development from 1959 onwards and the current development from 2006, yet the visitor is largely unaware of this post-liberation history. Ravensbrück's history as a Soviet army base, the second largest in the GDR, is also not part of the current representation. Furthermore, Ravensbrück's focus on the communist victim in the GDR and, most importantly, the future mothers, evident in Anna Seghers' phrase at the entrance to the memorial site and the various statues placed around the site, is an untapped resource for making visitors aware of the different memorialisation processes. Similarly, at Flossenbürg, the building of a Catholic chapel in 1946 as the only memorial could be the basis for discussions. Whilst such content does not engage visitors extensively with the Nazi past, it allows them to gain an understanding of the politicisation of history, which in itself has an educational function. At Ravensbrück, there is also the opportunity of illustrating the voices of the Soviet soldiers who subsequently lived onsite. Although this would require extensive historical research, it would offer Ravensbrück a unique additional opportunity to engage with the GDR legacy, in particular the complex relationship between the Soviet soldier and the local community. With this in mind, the memorial team could also integrate the voices of the curatorial team which was confronted with managing the site once the Soviet army left. The land was so contaminated that it could not be brought back into use without extensive restoration works, and Soviet army buildings changed the character of the historical landscape for good. Hence, the curatorial team had to and still has to make conflicting decisions about Ravensbrück's future presentation. Most buildings were constructed using cheap methodologies and therefore do not reflect modern building regulations (Eschebach 2020). Thus, opening previously neglected buildings always requires an adaption to provide safe access. By including the voices of curators and conservators, the visitor gains an understanding of the historical layers of the memorial site and their complex management.

6.5 Ongoing engagement

In 2019, the International Council for Museums (ICOM) caused a stir by introducing a new definition for museums. The proposed text defined museums as 'participatory and transparent' which should aim to contribute 'to human dignity and social justice, global equality and planetary wellbeing'. Particularly controversial was the statement describing museums as 'democratising, inclusive and polyphonic spaces for critical dialogue about the pasts and the futures', since this definition no longer contains the concepts of 'education' or 'collections'. Jette Sandahl, the director of ICOM who subsequently resigned over a row about this definition, clearly hit a raw nerve when she and her colleague proposed the democratisation of the museum. In fact, she argues that museums are not just mirrors of society, they are embedded within our society, and consequently the notion of a 'neutral museum' is an outdated concept (Sandahl 2020). Societies change rapidly and museums will have to learn to present controversial viewpoints and shifting perspectives.

Cornelia Siebeck (2014) notes that she has a feeling of discomfort whenever she reflects on the management of the German concentration camp memorials. The rapid institutionalisation of the memorial sites following the fall of the Berlin Wall was accompanied by the formation of expert commissions which influence not only the management of the sites but also the way they are represented. Much like their 'ordinary counterparts', memorial sites with their accompanying exhibitions are not neutral spaces. They are influenced by the societal structures and the individual members of staff who make decisions about what to preserve and what to exhibit. And, although Germany's *Gedenkstättenkonzept* values the contribution of civil initiatives, its strict funding criteria with a focus on 'scientific research' excludes lay projects. Consequently, memorial sites became scientific institutions that focused on the securing of historical traces and of historical facts. Outwardly, the memorial sites are now perfect with scientifically accurate exhibitions and predictable events, and in so doing lose the contact with society (Knoch 2015). In fact, Habbo Knoch (*ibid*), director for Lower Saxony's Memorial Foundation from 2007 to 2014, paints a gloomy picture by suggesting that German memorials do a lot but achieve little. He argues that the memorial sites do not take visitors out of the comfort zone, only hinting at violence, and as such present the past through a filter. The controversies surrounding the new exhibition at the House of the Wannsee Conference demonstrate this poignantly. All violent images were removed in order to avoid overburdening visitors and not to run the risk of a relative recognising a victim. Moreover, attempts to create a connection to present-day acts of violence are thwarted out of fear of trivialising the Holocaust.

Yet Nina Simon (2010) argues that visitors are disappointed with museums when the following participatory techniques are not addressed: the institution is irrelevant to my life; the institution never changes – I have seen it once; the authoritative voice of the museum does not include me; the institution does not allow me to express myself; the institution is not a comfortable place for me to talk to others. These experiences are present, more or less, at the German memorial sites, especially the sentiment that the organisation does not change. Habbo Knoch (2015) asks a very difficult question when he reflects about the future of the memorial sites: what happens if the moral lessons at the memorial sites turn to rejection and protest, precisely because they represent the reason of the state. This is indeed a risk that the memorial sites face if they continue to speak with the 'authoritative, scientific voice'.

My fieldwork in Bautzen illustrated the discrepancy between memorial-isation and day-to-day life. Whilst the memorial exhibits the human rights violations in the Stasi prison and the Soviet Special Camp (and now Nazi imprisonment), anti-immigration sentiments in town resulted in frequent clashes between asylum-seekers and the local population. If memorial sites want to remain relevant, then a different approach is required at times. Instead of expecting people to come to the memorial sites, memorials need to go to where the people are. One way to achieve this is by developing projects in schools rather than insisting that school visitors come to the site. Another way is the co-production of exhibition content which also avoids the top-down approach to exhibition design. For instance, at the Sixth District Museum in South Africa residents are invited to share their memories of living under apartheid (Bennett 2017). Those memories are included in an interactive map which means that the local community has a stake in the exhibition content. Another example of how the co-production of exhibitions can achieve a high level of engagement, especially with regard to challenging content, was Riga's exhibition on totalitarianism in the former KGB prison in 2014. The exhibi-tion explained the impact of the Stalinist and subsequent Soviet dictatorship on the Latvian population. Latvians who had emigrated to escape the dictatorship were asked to send in objects they had taken with them when they fled. These objects ranged from a bicycle to the starter culture for rye bread and were exhibited alongside the corresponding stories of the families. The exhibition not only engaged a vast Latvian diaspora across the world but has also explained a testing history in a very engaging way.

The Holocaust Museum Washington D.C. initiated a collaborative research project, *The Children of the Lodz Ghetto*, in which the aim was to collect infor-mation about the fate of the children (Simon 2010). The general public could join this project as volunteer researchers, and whilst only one third of the research information could be used as others were incorrect, the large engage-ment with the project surpassed the downsides. Such projects require the relin-quishing of power which the German memorial sites would find difficult, yet they can create long-lasting relationships with the local community, which is vital in politically charged situations such as in Bautzen. My case studies cur-rently do not consider the co-creation of content. And yet all four sites offer ample possibilities to include residents in exhibition design. For instance, in Bautzen, the memorial could explore the issues of having lived side by side with a Stasi prison or simply having lived under the GDR regime. It would add a new layer to the exhibition and initiate opportunities for breaking down bar-riers between the residents and the memorial site. In fact, it is vital to conduct such a project in order to avoid the alienation which occurred at Ravensbrück or other memorial sites. Annette Leo (2007), while conducting research on the memories of the perpetrator in Ravensbrück, discovered that the local com-munity was reluctant to answer any questions. They were afraid of the stigma, especially after the Ravensbrück supermarket scandal in 1991 (which was intended to be built on the edge of the former concentration camp). This fear has led to a division between the memorial and the local community which prevents historical research and ongoing engagement with the site. Hence, including the local community in co-production of exhibitions is important to create a sense of ownership and develop exhibitions that are relevant. It would also develop multivocality in the exhibition design, stressing how these sites were integrated within the wider community.

The sentiment 'I have already seen this elsewhere' was a commonly expressed viewpoint. It is a very difficult one as the consequence will be that visitors no longer engage with memorial sites. To counteract such developments, memorial sites need to consider ways of keeping the interest of potential visitors. Although Ravensbrück and Flossenbürg had temporary exhibitions during my fieldwork, they were rarely visited by people I accompanied. Indeed, the temporary exhibition at Flossenbürg consisted of a showcase of photographs of contemporary life in Israel that confused visitors. At Ravensbrück, the temporary exhibition explained the female medical staff of the Red Army and their destinies at the concentration camp. Temporary exhibitions will, however, always compete with the copious amounts of other information, especially at Ravensbrück which already has five exhibitions. Hence, temporary exhibitions are mainly about retaining the interest of repeat visitors or engaging residents. Nevertheless, temporary exhibitions which provide an in-depth analysis of specific aspects of the site can keep the interest of those visitors who would normally no longer visit the sites. As an example, Ravensbrück during the GDR era would open visitors' eyes to the politicisation of history and would at the same time address the bewilderment of East German visitors about the present-day exhibition concept. Flossenbürg already has a very high percentage of repeat visitors and could therefore capitalise on it by providing site-specific temporary exhibitions. For instance, Flossenbürg was the hub of an extensive network of 90 subcamps reaching as far as Saxony and today's Czech Republic. Visitor reactions were often 'I never knew about these subcamps'. Highlighting the story of these subcamps in detailed temporary exhibitions would enhance the understanding of the great army of unknown victims (Frankl 1946) whilst creating a link to Flossenbürg. Bautzen's unique position as a Stasi prison that included West German prisoners (once they were sentenced) could facilitate a fascinating exhibition explaining their destinies, including the voices of the staff who had to deal with these prisoners. This is certainly an aspect which is not known, and judging from the response of visitors when I conducted the interview with an eyewitness, the interest in such an exhibition would be high. Bautzen could also initiate relationships with memorial sites in other countries of the former Eastern bloc. By exchanging exhibitions between different nations visitors could develop a greater sense of the communist dictatorships in the former Eastern Europe. How temporary exhibitions can be successfully linked to contemporary issues was shown by the Human Rights Centre in Cottbus, located in a former GDR prison. Cottbus, like Bautzen, has experienced far-right protests and has therefore decided to open an exhibition about migrants in the GDR. Under the title *Prescribed Solidarity – the Treatment of 'Foreigners'* in the GDR, Cottbus exhibited the behaviour towards migrants in the GDR, an unknown part of GDR history. At the same time, it opened up the debates about current far-right sentiments in the city, thus retaining visitors' interest by creating a link to the present. At the House of the Wannsee Conference, the display of the interpretation panel about the Evian Conference in 1938 sparked lively debate. It demonstrates that forging links with current political situations encourages visitors to think beyond the displays. Ravensbrück could engage with the suffering of women in more recent conflicts such as Srebrenica, Myanmar or Rwanda. Within the German context, including current conflicts in exhibitions about the Nazi past is often considered inappropriate as 'one cannot compare'. I do not suggest comparisons between the Nazi past and more recent conflicts; however, addressing those situations will emphasise that we are still confronted with severe violence which will hopefully make visitors think critically, an aim the memorial sites surely should have. And as Wagner (2017) suggests, the victim-centric focus at memorial sites needs to change by considering the bystanders, onlookers and perpetrators. In so doing, one would need to address issues such as indoctrination, racism, antisemitism or fear, all subjects which are relevant for present-day conflicts and not just applicable in the context of National Socialism.

In 2016, the Bergen-Belsen Memorial offered a photography workshop under the theme *Was bleibt* (What remains). Participants learned how to compose photographs with DSLR (digital single-lens reflex) cameras while at the same time exploring the historical significance of the site. They were encouraged to take photographs of places that resonated with them. When Shanti Sumartojo (2019) asked visitors at Camp de Milles to take photographs of places which made an impact on them, a whole range of images emerged that revealed the emotional connection between the visitor and the site. It also engages visitors on a different level as they have to look more closely at the historical layers in front of them. Donald Mitchell (2000) reminds us that we do not read the landscape in the same way since we attach different meanings to them. Thus, connecting visitors to a landscape by using creative approaches can engage visitors who might otherwise not consider visiting a memorial site and/or think they have seen so many memorial sites that they no longer want to visit them. Another method to engage with visitors is the Imperial War Museum's initiative of conflict cafés, allowing visitors to meet experts in a relaxed setting. For instance, the Imperial War Museum North organised a conflict café on 'the war in Syria', or in London, the Imperial War Museum's conflict café addressed 'life after terrorism', focusing on the recent attacks. This is an informal space for visitors to explore issues and join discussions. These debates are facilitated by curators, academics, journalists or artists and provide multiple perspectives. In Bautzen II, this method could engage visitors with those pertinent questions they might have about the GDR and could therefore contribute to the wider *Aufarbeitung* (investigation into the past) of the GDR legacy in contemporary Germany. At the concentration camp memorials, it would provide a space for visitors to engage with the Nazi past and often with their own family memories.

The visitor research has also shown that visitors take children to memorial sites, even under the recommended age guidelines. Yet the memorial sites do not cater for individual child visitors. In particular, at the House of the Wannsee Conference it resulted in children being uninterested and switching off, consequently impacting the whole visit. If visitors decide to travel to a memorial site to explain this history to their children, then these visits should be viewed as educational possibilities instead of restricting them. At the Nazi documentation centre in Cologne, staff provided an opportunity specifically for children to ask questions about fascism that were answered by curators. By doing so, the documentation centre created links with the community and, most importantly, offered interested families the chance to engage with the most pertinent issues of our time. Hadamar memorial site in Germany, one of the locations of Hitler's euthanasia programme, currently leads in terms of engagement with children. They use the biographies of children murdered at Hadamar as an age-appropriate entry point into the history of the Nazi period. Children are encouraged to use musical instruments to express the feelings of the child victims or take part in theatre productions. Whilst evaluation of such programmes is limited, the educators in the programme highlight that children frequently asked when the next project would take place (Gabriel 2008). Thus, the question should not be *whether* children can visit a memorial site, the question should be *how* they can be enabled to visit such sites (*ibid*). Ravensbrück, which had approximately 880 child prisoners, could find ways to integrate the biographies of some of those children by focusing on their lives in the camp. These stories will engage the family visitor and provide parents with the opportunity to engage their children, a wish expressed by visitors not only to Ravensbrück but also to the House of the Wannsee Conference.

Memorial sites also rarely offer a social encounter (Arnold-de Simine 2013). My visible presence at the memorial sites has also shown that there is a desire for some visitors either to share their experiences or to ask questions. None of my case studies currently has a staff presence which would allow visitors to have these conversations. The change from an authoritative exhibition style to a communicative one would go hand in hand with the presence of staff, especially if we pose open-ended questions. It is often those face-to-face conversations that provide the most long-lasting impressions. This staff presence is particularly important at Bautzen II where the legacy of the GDR is still in living memory and where visitors wrestle with the impact of the Stasi oppression. Bautzen has revealed that a strong auratic atmosphere can overwhelm visitors. Emotions run high at times, and whilst there is little the memorial can do to regulate this, they need to be aware of it. Creating spaces where visitors can take a break, and as suggested earlier, providing a member of staff on the ground, might help to mitigate these overwhelming feelings.

I am aware that my suggestions might be an unachievable utopia, especially within the German context. Collaborative projects would require a fundamental change in how memorial sites are managed. In addition, current staffing levels would not allow the German sites to start outreach projects and/or increase the number of events onsite. Yet Germany's governance structure excludes the individual visitor from the outset, which can be addressed by changing organisational cultures.

6.6 Governance of memorial sites

Since 1999, the federal government of Germany has institutionalised memory by including 'exemplary' memorial sites into the overarching memorial concept. To qualify for inclusion into the federal scheme, the memorial sites have to fulfil a set of criteria: be of national and international importance, be an historically authentic site and be representative of the persecution under the Nazi regime or the GDR dictatorship (Bundesregierung Online 2015). In addition, aside from the memorial sites being places of commemorations, they also have a significant role as places of learning within the current societal context. As such, it is essential to offer a wide-reaching educational programme. Thus, the German government considers the sites as active participants in political education. This institutionalisation has secured the financial future of the sites, and they do not have to worry about achieving income targets, unlike, for instance, the Terezin memorial which relies on external funding alongside state funding. The key German memorial sites are managed by *Stiftungen* which are comparable to foundations in the United Kingdom. They are supported financially by the federal government and the regional governments. Unlike charities or non-profit organisations, foundations work towards one specific aim. In the context of the German memorial sites, the *Brandenburgische Gedenkstätten* state as their purpose:

> The purpose of the foundation is to commemorate terror, war and tyranny, to promote the public's engagement with the subject and to allow a dignified commemoration of the victims of the tyranny of the Nazi regime, the Soviet occupying forces and the GDR.

The aim of the *Bayerische Gedenkstätten* is as follows:

> The purpose of the foundation is to preserve and design the memorial sites as witnesses of the crimes of National Socialism, as places of remembrance of the suffering of the victims and as learning places for future generations, to support the related historical research and to ensure that the knowledge about the historical event is kept alive and carried on in the consciousness of the people.

And *Sächsische Gedenkstätten's* mission statement is:

> The purpose of the foundation is to develop, promote and supervise those sites in the Free State of Saxony, that are authentic places of political violent crimes of international importance, of special historical significance, reminiscent of political persecution, state terror and state organised murders. They develop these sites as places of extracurricular education as well as political education in the European context. The foundation has to honor the victims of the National Socialist and the communist dictatorships, in particular the SED dictatorship, resistance against these dictatorships and its structures, and to document methods of the respective power systems for the public.

Hence, all three foundations emphasise the importance of public engagement and education. It is therefore somewhat surprising that Matthias Heyl and Heide Schölhorrn stated in 2007, while reflecting on the newly opened perpetrator exhibition in Ravensbrück, that the pedagogical team was unable to provide an insight into the reception of this exhibition amongst the adult audience as they are not included in the overall educational offer. A decade later, we are in much the same position. All four memorial sites do not currently evaluate the success of their exhibitions and/or educational offers with regard to individual visitors. In fact, educational offers for individual visitors do not exist, apart from the occasional guided tour.

The financial security the German sites enjoy has caused a sense of complacency. This has also resulted in a structure where accountability is not essential. For instance, the recently opened Museum of Peace and Justice in Montgomery/USA is largely financed by donations from private donors and grants (75 per cent). With the reliance on grants and donations comes the responsibility for evaluating charitable outcomes. Whilst I do not suggest subjecting German memorial sites to financial hardship, there ought to be a culture which fosters accountability. Jochen Fuchs (2019), for instance, argues that since the conference about evaluations at German memorial sites in 2003, very little has changed. Even basic statistics about visitor demographics are hard to find.

In the case of Hohenschönhausen, Kowalczuk (2018) argued that the memorial site became the director's personal project. Indeed, the success of Hohenschönhausen and other memorials is largely measured on visitor numbers. Yet visitor numbers are not an indication of whether the exhibition is successful. No other site highlights this fact more than the House of the Wannsee Conference. It proudly announces its high visitor numbers every year. However, while based at the reception for four weeks, I found that a large proportion of visitors arrive as part of a 'Holocaust tour'. These visitors often do not engage with the exhibition as the memorial is one stop on a busy tour operator's itinerary. After an introduction by the tour guide, they tend to visit the memorial's exhibition rooms briefly, with a strong focus on the former conference room itself, before heading back to the coach.

The lack of understanding on how to manage memorial sites more sustainably is not unique to Germany. The director of the newly founded research group Economics at Memorial Sites at the SWPS University for the Social Sciences and Humanities (Warsaw, Poland) pointed out that research has not been conducted on how to manage these sites, with ever-rising conservation costs and the expense of modern technology in exhibition design (Sawicki 2018). Ensuring financial security, whilst also meeting demands of a new generation of visitors, is therefore an under-researched area. For Günter Morsch (2018), the former head of *Brandenburgische Gedenkstätten*, it is predominantly the lack of financial resources that prevent the transformation of the memorial sites into modern museums. I argue it is also the organisational culture.

Knabe's dismissal at Hohenschönhausen revealed a series of structural failures within the German memorial landscape. Knabe was criticised for centralising power in the director's office. This meant that employees were not participants in the future development of the memorial. Hohenschönhausen is, nevertheless, no exception. Matthias Heyl (2016) pointed out to me in a conversation that the educational team was not part of the discussion around the new exhibition in Ravensbrück. Furthermore, when I conducted my talk about the visitor research results for the staff team at Ravensbrück, the front-of-house team informed me afterwards that this was the first time that the whole team had taken part in an event. This situation is very similar to the one in Flossenbürg where a distinction is made between the experts and the non-experts, with the 'non-experts' being instructed not to answer any questions. In essence, the front-of-house team, outsourced to an external security company, could not engage with visitors since they did not possess the historical knowledge, and yet they were the only visible members of staff for individual visitors. In Bautzen II, there was an obvious disagreement between the local management team and the headquarters of the foundation in Dresden. The local team felt stifled by the headquarters in trialling new visitor initiatives. Whilst such examples can be found in any larger organisations, at memorial sites it is particularly detrimental. It has a direct impact on the visitor experience.

A visitor I accompanied complained that Ravensbrück was closed on a Monday (unofficially it was open, however). The response from the security staff was: 'The camp has been closed since 1945.' This rather cynical answer is clearly not an aim the memorial sites should aspire to when addressing individual visitors. Similarly, at the House of the Wannsee Conference, I experienced the closure of the actual conference room due to large school groups (for very lengthy periods) or the sudden entering of a room with a large group, so that the visitor was left only with the option to leave. In these cases, no conversation took place with the visitor. In Flossenbürg, a tour guide suggested that the 'Nazi evil is in the German DNA'. To take such an approach with individual visitors cannot be the educational aim of a memorial site.

Charles Handy (1976), the leading expert on organisational culture, said that if you want to change an organisation, change its culture. If the memorial sites are required to play an essential part in political education, then one cannot create an internal structure where the individual visitor is neither heard nor seen. The current management structures consist of curatorial and pedagogical teams, so the only visitor-facing department is the museum education team, and their focus is the school visitor. The individual visitor does not feature. At Auschwitz, for instance, there is a different approach to visitor management. Admittedly, it is a large site; nevertheless, departments such as visitor services, methodology of guiding, educational projects or e-learning clearly focus on individual visitors, and crucially, they consider the future. Re-structuring the management at German memorial sites incorporating a department 'Learning and Participation' would create a new structure that includes individual visitors. I deliberately used the term 'participation' as it moves away from the top-down approach and encourages staff to think about engagement rather than about strict pedagogical guidelines. Participation is also much wider than education as it includes informal learning opportunities for visitors that are beyond the national school curriculum. Yet such an approach requires a monumental shift in managing memorial sites, away from the centralisation of power to one which draws on the expertise of its employees and external stakeholders.

Aside from the management team, there are expert commissions which influence exhibitions. These expert commissions include various representatives of victim associations and historians. In the past, victim associations had a major stake in the content of the exhibitions, and rightly so. For the surviving victims, the memorial sites were first and foremost a cemetery where they could remember not only their own suffering but also that of the victims who did not survive. In addition, this allowed them to have their voices heard. Within the new phase without eyewitnesses, memorial sites are not just places of evidence anymore. Their function is changing, with an increasing desire for political education as an outcome of the visit. This requires a different approach to exhibition design and therefore needs to include experts with a wide range of skills such as education, psychology and principles of design or accessibility (writing of content). For instance, Gerd Kühling (2017), the designer of the new temporary exhibition *Repressed Memory – How West Berlin Dealt with Sites of Nazi Perpetrators*, explained that the content was written using an easily understandable German with the aim of engaging a wide audience. At the end of the exhibition, visitors were asked to share their thoughts on the following questions: What is still hidden? What is missing in the current culture of remembrance? What other topics should memorial sites address? Which new approaches would you suggest in dealing with the Nazi regime? Equally, the memorial site of Bergen-Belsen asked its social media followers recently 'What do you expect from us?' A different thinking seems to be appearing in Germany, but it is slow.

In summary, I therefore suggest the following mission statement for German memorial sites with a view to remain relevant for future generations:

> The memorial site is a living reminder of the human suffering in the hands of the Nazi or SED dictatorships. The conservation of historical evidence

alongside ongoing historical research will inform the current and future presentation of the site, exhibition design, education and public engagement. As such, the memorial site will, in collaboration with national and international stakeholders, design programmes that encourage critical debate and lifelong learning, promote the principles of democracy and human rights, and stimulate self-reflection during a visit onsite.

6.7 Where next?

Whilst this research project has provided insights into the visitor behaviour at memorial sites, it has simultaneously opened up a range of new questions. Although society firmly believes that memorial sites are crucial for 'learning the lesson', we have very little knowledge about how visitors process these visits. If visitors display emotional reactions at memorial sites, what will they do with these emotions after the visit? Does this emotional response encourage critical thinking or even action, as Landsberg (2004) suggests? If the desire of the memorial sites is to create a long-lasting impact, then we need to conduct long-term studies to measure the effect of such visits on the visitor. Furthermore, I was also able to concentrate only on the overarching impact of exhibition design and could not consider certain aspects in detail. Sexual violence is one of these sensitive topics that require further exploration of exhibition practice and visitor response. The section on sex slave labour at Ravensbrück resulted in two different responses, either in intense engagement or in leaving quickly. It is a highly delicate topic, not least due to stigma; however, with an increased awareness in contemporary societies about women's experiences of violence, it is an area that deserves attention. We need to understand how we can exhibit such sensitive topics without overwhelming the visitor and respecting the dignity of the victim. Further research might even reveal that such exhibits should not form part of an exhibition at a memorial site and are better left to other forms of education, for example, documentaries.

The research has also shown that visitors desire an emotional connection with the site. This is a fine balance which can easily change into emotional exhaustion. But rather than making assumptions about possible visitor reactions, we need to consider cross-disciplinary research, for instance with researchers in psychology, to understand the emotional reaction of visitors and the potential long-term consequences. Psychological research would also enable us to examine the 'true' underlying emotions when visiting memorial sites and thus reduce the risk of cultural contamination by visitors mentioning certain emotions due to a societal expectation.

A conversation with the manager at Bautzen II, Silke Klewin, emphasised that visitors to Bautzen II are often familiar with the history of the GDR and the human rights violations. However, the visitors with whom she would like to engage, for example, the people who attend far-right demonstrations, do not come. Sachsenhausen hosted a group from the German far-right political party AfD, sparking criticism about hosting such groups at memorial sites.

Hence, there is a debate about the 'political correctness' of visitors or who should (or should not) come to a memorial site. It is however exactly those visitors with whom we need to engage rather than excluding them. We also need to understand why people decide not to visit a memorial site or never even consider doing so. It is common practice in museums to conduct non-visitor studies in order to understand barriers, but these studies do not take place at memorial sites. If we want memorial sites to have a 'transformative' function, then we need to understand why people choose not to visit a site. If barriers can be identified, it could open up new lines of enquiry for the role of memorials and exhibition designs.

I have shown in my analysis that the research in dark tourism insufficiently explains the phenomenon of tourism to 'dark sites'. There is nevertheless one aspect which my research also neglected: the local community. The impact memorial sites have on local communities is underestimated. Jody Manning (2010) emphasised in his research how the local community in Oświęcim (Auschwitz) is often accused of being inconsiderate simply because they had a wedding or held a party in a restaurant. At Flossenbürg, visitors were shocked to find the display of the German flag at houses near the memorial site during the European football championship which coincided with my fieldwork. In Bautzen, residents want to move on from the legacy of the Stasi and not be confronted with a daily reminder of the GDR dictatorship. In Ravensbrück, the development of a supermarket in 1991 on the land of the concentration camp (but away from the actual camp) caused an international outcry but very little concern amongst the local community, including the town council. Without international pressure this project would have been completed, so in many ways decision-making processes were taken out of the hands of the local community. So far, only two major German studies, Leo and Grätz's (2008) *'Das ist so'n zweischneidiges Schwert hier unser KZ . . .': Der Fürstenberger Alltag und das Frauenkonzentrationslager Ravensbrück* ('It's such a dou-ble-edged sword here, our concentration camp . . . .' Daily life in Fürstenberg and the Ravensbrück women's concentration camp) and Lena Möller's (2019) *'Auf Stätten des Leids Heime des Glücks': Die Siedlung am Vogelherd auf dem Areal des ehemaligen KZ Flossenbürg und ihre Emotionalisierung als Wohn- und Gedächtnisort* ('On sites of suffering homes of happiness': The Vogelherd estate on the site of the former Flossenbürg Concentration Camp and its emotionalisation as a place of residence and memory) currently address the relationship between the memorial site and the local community. In particular, Möller's (*ibid*) work shows the ambivalent feelings of the local residents in light of the increased focus of Flossenbürg as a memorial site. Ever–new initiatives for memorials and increasing visitor numbers mean that we cannot ignore the impact these sites have on the local population. In the end, as a visitor I can leave the site; in most cases, the local community cannot unless they move house. We therefore need to conduct research into the implications of memorials for residents, especially psychologically, with a view to developing supportive relationships between the visitor and the host community.

As a generation of eyewitnesses of the Holocaust dies, there are endless new initiatives to preserve the memories of these witnesses. One of those developments is the 'virtual Holocaust survivor' using artificial intelligence. The New Dimensions in Testimony project, a collaboration between the University of Southern California (USC) Institute of Creative Technologies, the USC Shoah Foundation and Conscience Display, created a 3D prototype of a Holocaust survivor. The most common questions were recorded and then converted into a hologram (USC Institute for Creative Technologies 2017). The hologram will 'interact' with future visitors; people will be able to ask questions and the hologram will respond, much like Microsoft's Cortana or Amazon's Alexa. Another development is *The Faces of Auschwitz* project. A collaboration among the Auschwitz-Birkenau Museum, a Brazilian colourisation studio, academics and journalists aims to colourise registration photographs. The idea is that photographs in colour will bring the story to life, preserve the memory as a warning for future generations and engender empathy amongst visitors.

The Covid-19 pandemic has accelerated the process of digitisation. Since school excursions to memorial sites and events at annual anniversaries were no longer possible, the German memorial sites, having been previously very sceptical about these developments, caught up by introducing a range of 'virtual tours'. Ravensbrück and Sachsenhausen published an animation film, while Dachau created an audio-visual guided about the liberation of the camp. In many ways Bergen-Belsen was a trailblazer when it launched an augmented reality app in 2015 that shows visitors how Bergen-Belsen used to look. Jens-Christian Wagner (2018) argues that augmented reality can be an important educational tool for places with few remains; yet applications that insinuate an 'authentic place' and/or reduce the multi-perspectivity of the sites should be avoided. The difficulty with such suggestions is that we crossed the threshold of 'virtual simulations of authentic sites' some time ago, and German memorial managers will not be able to halt the development. Christian Kuchler (2021) rightly notes that we are on the cusp of a new era that is as significant as the broadcasting of the American *Holocaust* TV Series in 1979. These new applications 'bring history to life' in better ways than the 'real thing'. In 2017, the Italian design studio 101 introduced a virtual reality simulation of Auschwitz at the Gamescon in Cologne, which aims to show daily life in the camp. It, however, leaves one crucial aspect out: violence (Jong 2020). At the beginning of the simulation an invisible hand with a suitcase emerges, showing the entrance to Auschwitz on a bright summer's day. The suitcase then dissolves, the site turns dark and the sounds of singing birds are replaced with the sounds of sirens and Germans shouting commands. The game was met with intense rejection, criticising the manipulation of history and stating ethical concerns. I am reminded of Umberto Eco's (1987) essay *Travels in Hyper Reality* which he wrote in 1987, criticising simulacra such as Disneyland in the United States. He could not have foreseen that his essay would be even more relevant in the 21st century. Eco argued that the danger with simulacra is that they can become more real than the reality. Bearing in mind, that visitors in this research project already mentioned the lack of an 'aura' at the concentration camp memorials, which impact will these perfect simulations have on future visitor behaviour? Moreover, it is apparent that most of the projects focus on Auschwitz. Considering the research results of this project and the already dominant narrative of Auschwitz, how will this impact the memorial sites in Germany? What about the memorial sites which do not have the financial resources to implement these schemes: will their memories fade, thus starting another episode of selective memory?

Kuchler (2021) conducted research with two German school groups, which tested the student responses to the film *Inside Auschwitz: Das ehemalige Konzentrationslager in 360°*. One group of students had been to Auschwitz before, while the other only knew Auschwitz through media representations. Although the research was hampered by the pandemic resulting in a small sample size, the initial findings should make us think. The group who had already visited Auschwitz reacted more emotionally than those who had not visited Auschwitz before. In fact, the group with no physical experience of Auschwitz felt overwhelmed by the virtual reality (VR) simulation as the images overlapped each other, and they did not know what to focus on: the images or the eyewitness accounts. Thus, Kuchler argues (*ibid*) that a virtual visit cannot replace a physical visit. For me, digital applications have other significant downsides that emerge out of this research. Visitors connect to memorial sites through their senses which they value highly. With a tablet in my hand showing the landscape as it used to be and providing me with additional historical context, I encounter the landscape through a screen. Indeed, the screen could act as a barrier to the emotional connection visitors seek. I am, of course, aware that my suggestions mentioned earlier also include the use of digital technology, yet my aim is to bring visitors closer to the landscape that is in front of them rather than encouraging them to engage with the landscape via a screen. Nevertheless, I agree with Sumartojo (2020), who argues that virtual encounters inevitably transform the way visitors approach the sensory and material aspects of the site which needs to be understood.

Hence, I share the sentiment of Volkhard Knigge, who once said that he would like to close Buchenwald Concentration Camp memorial for a year for 'thinking' (Greves 1999). The visitor research has shown that we are at a crucial moment in time where serious questions about the presentation need to be asked. In fact, we need to distinguish between memory and learning, two fundamentally different concepts that have in recent years collapsed into each other with ever-new initiatives for memorialisation and a belief system that a memorial site can 'transform' an individual. Thinking needs to be the basis of the memorialisation processes, and by doing so, we might one day achieve 'Never Again'.

6.8 Bibliography

Adams, Tim. 2016. 'Neil MacGregor: "Britain Forgets Its Past: Germany Confronts It"'. *The Guardian*, 17 April, sec. Culture. www.theguardian.com/culture/2016/apr/17/neil-macgregor-britain-germany-humboldt-forum-berlin.

Alexander, Jeffrey C. 2002. 'On the Social Construction of Moral Universals: The "Holocaust" from War Crime to Trauma Drama'. *European Journal of Social Theory* 5 (1): 5–85.

Amjahid, Mohamed. 2021. 'Holocaustgedenken: Die deutsche Erinnerungsüberlegenheit'. *www.spiegel.de*. www.spiegel.de/kultur/holocaust-gedenken-die-deutsche-erinnerungsueberlegenheit-a-056d10a7-2b3c-4383-804e-c2130ed6581d.

Arnold-de Simine, Silke. 2013. *Mediating Memory in the Museum: Trauma, Empathy, Nostalgia.* Basingstoke: Palgrave MacMillan.

Assmann, Aleida. 2011. *Erinnerungsräume: Formen und Wandlungen des kulturellen Gedächtnisses.* 5th ed. München: C.H. Beck.

Assmann, Aleida, and Juliane Brauer. 2011. 'Bilder, Gefühle, Erwartungen. Über die emotionale Dimension von Gedenkstätten und den Umgang von Jugendlichen mit dem Holocaust'. *Geschichte und Gesellschaft* 37 (1): 72–103.

Behrens-Cobet, Heidi. 1998. 'Erwachsene in Gedenkstätten – Randständige Addressaten. Zur Einführung'. In *Bilden und Gedenken. Erwachsenenbildung in Gedenkstätten und an Gedächtnisorten*, edited by Heidi Behrens-Cobet, 7–22. Essen: Klartext-Verlag.

Bennett, Bonita. 2017. 'District Six Museum: Activists for Change'. *Museum International* 68 (3–4): 5–10. https://doi.org/10.1111/muse.12138.

Bertelsmann Stiftung, Jana Faus, Matthias Hartl, and Kai Unzicker. 2020. *30 Jahre Deutsche Einheit. Gesellschaftlicher Zusammenhalt im Vereinten Deutschland.* Gütersloh: Bertelsmann Stiftung. www.bertelsmann-stiftung.de/en/publications/publication/did/30-jahre-deutsche-einheit-all.

Bormann, Natalie. 2018. *The Ethics of Teaching at Sites of Violence and Trauma: Student Encounters with the Holocaust.* New York: Palgrave Macmillan.

Breithaupt, Fritz. 2019. *The Dark Sides of Empathy.* Translated by Andrew B. B. Hamilton. Ithaca; London: Cornell University Press.

Bruha, Toni, Maria Berner, Herma Löwenstein, Anna Poskocil, Anna Schefzik, Hermine Huber, Irma Trksak, et al. 1945. *Frauen-Konzentrationslager Ravensbrück. Geschildert von Ravensbrücker Häftlingen.* Wien: Stern-Verlag.

Bundesregierung Online. 2015. 'Fortschreibung der Gedenkstättenkonzeption'. www.bundesregierung.de/Content/DE/_Anlagen/BKM/2008-06-18-fortschreibung-gedenkstaettenkonzepion-barrierefrei.pdf;jsessionid=5EB1259FE3B3A61EB2675A5898064E8F.s4t1?__blob=publicationFile&v=3.

Charlesworth, Andrew, and Michael Addis. 2002. 'Memorialization and the Ecological Landscapes of Holocaust Sites: The Cases of Plaszow and Auschwitz-Birkenau'. *Landscape Research* 27 (3): 229–251.

Creţan, Remus, Duncan Light, Steven Richards, and Andreea-Mihaela Dunca. 2019. 'Encountering the Victims of Romanian Communism: Young People and Empathy in a Memorial Museum'. *Eurasian Geography and Economics* 59 (5–6): 632–656.

Eberle, Annette. 2004. 'Pädagogik als Projekt'. *Gedenkstättenrundbrief* 122: 13–24.

Eco, Umberto. 1987. *Travels in Hyper Reality: Essays.* London: Pan Books in association with Secker & Warburg.

Erpel, Simone, and Insa Eschebach. 2020. 'Im Gefolge der SS: Aufseherinnen des Frauen-Konzentrationslagers Ravensbrück. Konzeption und Geschichte eines Ausstellungsprojektes'. *Gedenkstättenrundbrief* 200: 18–31.

Eschebach, Insa. 2020. 'Die Mahn und Gedenkstätte Ravensbrück Planungs- und baugeschichtliche Entwicklungen'. In *Gestaltete Erinnerung: 25 Jahre Bauen in der Stiftung Brandenburgische Gedenkstätten 1993–2018. Eine Dokumentation*, edited by Günter Morsch and Horst Seferens, 261–283. Berlin: Metropol Verlag.

Ette, Wolfram. 2019. 'DDR – Wir haben keine Lösungen'. www.freitag.de/autoren/der-freitag/wir-haben-keine-loesungen.

Faulenbach, Bernd. 2019. 'Eine neue Erinnerungskultur? – Entwicklungslinien und Probleme der Gedenkstätten seit der Epochenwende 1989/90'. *Sachsenhausen Lectures* 3 (1): 1–44.

Foster, Stuart J., Alice Pettigrew, Andy Pearce, Rebecca Hale, Adrian Burgess, Paul Salmons, and Ruth-Anne Lenga. 2016. *What Do Students Know and Understand about the Holocaust?: Evidence from English Secondary Schools*. London: University College London.

Frankl, Viktor E. 1946. *Ein Psychologe erlebt das KZ*. Wien: Verlag für Jugend und Volk.

Fuchs, Jochen. 2019. *Auschwitz als eine moralische Anstalt betrachtet oder Was kann eine gute Gedenkstätte eigentlich wirklich bewirken*. Halle (Saale): Mitteldeutscher Verlag.

Gabriel, Regine. 2008. '10 Jahre Arbeiten mit Kindern an der "Euthanasie"- Gedenkstätte Hadamar'. *Gedenkstättenrundbrief* 154: 29–30.

Gensburger, Sarah. 2019. 'Visiting History, Witnessing Memory: A Study of a Holocaust Exhibition in Paris in 2012'. *Memory Studies* 12 (6): 630–645.

Giesecke, Dana, and Harald Welzer. 2012. *Das Menschenmögliche: zur Renovierung der deutschen Erinnerungskultur*. Hamburg: Edition Körber.

Giordano, Alberto, and Tim Cole. 2018. 'The Limits of GIS: Towards a GIS of Place'. *Transactions in GIS* 22: 664–676.

Greves, Andreas. 1999. 'Weinen bildet nicht'. *Spiegel Spezial*, 1 August. http://magazin.spiegel.de/EpubDelivery/spiegel/pdf/14104053.

Grindheim, Jan Erik. 2019. 'Why Right-Leaning Populism Has Grown in the Most Advanced Liberal Democracies of Europe'. *The Political Quarterly* 90 (4): 757–771.

Gryglewski, Elke, Verena Haug, Gottfried Kößler, Thomas Lutz, and Christa Schikorra, eds. Gedenkstättenpädagogik: Kontext, Theorie und Praxis der Bildungsarbeit zu NS-Verbrechen. Berlin: Metropol Verlag.

Handy, Charles B. 1976. 'So You Want to Change Your Organisation? Then First Identify Its Culture'. *Management Education and Development* 7: 67–84.

Halbwachs, Maurice. 1925. *Les Cadres Sociaux de La Mémoire*. Paris: Felix Alcan.

Hesselmann, Markus. 2007. 'Eine schräge Geschichte'. *Der Tagesspiegel*, 18 July. www.tagesspiegel.de/politik/geschichte/erinnerungskultur-eine-schraege-geschichte/989804.html.

Heyl, Matthias. 2016. Personal conversation. *Exhibition Design at Ravensbrück*, 20 July 2016.

Heyl, Matthias, and Heide Schölhorrn. 2007. 'Zur Auseinandersetzung mit Täterschaft in der Arbeit der pädagogischen Dienste der Mahn- und Gedenkstätte Ravensbrück'. In *"Im Gefolge der SS": Aufseherinnen des Frauen-KZ Ravensbrück: Begleitband zur Ausstellung*, edited by Simone Erpel, Johannes Schwartz, Jeanette Toussaint, and Lavern Wolfram. Berlin: Metropol-Verlag.

High, Steven. 2013. 'Embodied Ways of Listening: Oral History, Genocide and the Audio Tour'. *Anthropologica* 55 (1): 73–85.

Hirsch, Marianne. 2001. 'Surviving Images: Holocaust Photographs and the Work of Postmemory'. *The Yale Journal of Criticism* 14 (1): 5–37.

———. 2008. 'The Generation of Postmemory'. *Poetics Today* 29 (1): 103–128.

Jeismann, Michael. 2001. *Auf Wiedersehen Gestern: Die deutsche Vergangenheit und die Politik von morgen*. Stuttgart: Deutsche Verlags-Anstalt.

Jones, Sara. 2014. *The Media of Testimony: Remembering the East German Stasi in the Berlin Republic*. Basingstoke: Palgrave MacMillan.

Jong, Steffi de. 2020. 'Witness Auschwitz? How VR Is Changing Testimony'. *Public History Weekly* 8 (4). https://public-history-weekly.degruyter.com/8-2020-4/witness-auschwitz-vr/.

Klüger, Ruth. 2001. *Still Alive: A Holocaust Girlhood Remembered*. New York: Feminist Press at the City University of New York.

Knoch, Habbo. 2015. 'Wohin Gedenkstätten?'. *Gedenkstättenrundbrief* 178: 3–8.

———. 2020. *Geschichte in Gedenkstätten: Theorie – Praxis – Berufsfelder.* Tübingen: UTB GmbH.

Köpping, Petra. 2018. *'Integriert doch erst mal uns!' – Eine Streitschrift für den Osten.* Berlin: Ch. Links Verlag.

Kowalczuk, Ilko-Sascha. 2018. 'Und was hast du bis 1989 getan?'. *Süddeutsche Zeitung*, 23 October. www.sueddeutsche.de/kultur/ddr-geschichte-aufarbeitung-1.4179958?reduced=true.

Kuchler, Christian. 2021. *Lernort Auschwitz: Geschichte und Rezeption schulischer Gedenkstättenfahrten 1980–2019.* Göttingen: Wallstein Verlag GmbH.

Kühling, Gerd. 2017. '"Ausgeblendet. der Umgang mit NS-Täterorten in West-Berlin" Rede zur Ausstellungseröffnung am 20. Januar 2017 in der Gedenk- und Bildungsstätte Haus der Wannsee-Konferenz'. *Mitgliederrundbrief:* 7–11.

LaCapra, Dominick. 1994. *Representing the Holocaust: History, Theory, Trauma.* Ithaca: Cornell University Press.

Landsberg, Alison. 2004. *Prosthetic Memory: The Transformation of American Remembrance in the Age of Mass Culture.* New York: Columbia University Press.

———. 2009. 'Memory, Empathy, and the Politics of Identification'. *International Journal of Politics, Culture, and Society* 22 (2): 221–229.

Lehmann, René. 2011. '"Jetzt bist de ein zweites Mal betrogen worden!" – Vergleichendes Erinnern gesellschaftlicher Verhältnisse'. In *Soziale Gedächtnisse. Selektivitäten in Erinnerungen an die Zeit des Nationalsozialismus*, edited by Gerd Sebald, René Lehmann, Monika Malinowska, Florian Öchsner, Christian Brunnert, and Johanna Frohnhöfer, 43–66. Bielefeld: Transcript Verlag.

Leo, Annette. 2007. 'Die Aufseherin von nebenan. Gespräche mit Fürstenberger Bürgerinnen und Bürgern'. In *'Im Gefolge der SS': Aufseherinnen des Frauen-KZ Ravensbrück: Begleitband zur Ausstellung*, edited by Simone Erpel, Johannes Schwartz, Jeanette Toussaint, and Lavern Wolfram, 329–338. Berlin: Metropol-Verlag.

Leo, Annette, and Peter Grätz. 2008. *Das ist so'n zweischneidiges Schwert hier unser KZ . . .: Der Fürstenberger Alltag und das Frauenkonzentrationslager Ravensbrück.* 2nd ed. Berlin: Metropol-Verlag.

Maier, Charles S. 2002. 'Hot Memory . . . Cold Memory: On the Political Half-Life of Fascist and Communist Memory'. *IWM* (blog). www.iwm.at/transit/transit-online/hot-memory-cold-memory-on-the-political-half-life-of-fascist-and-communist-memory/.

Manning, Jody Russell. 2010. 'The Palimpsest of Memory: Auschwitz and Oświęcim'. *Holocaust Studies* 16 (1–2): 229–256.

Massey, Doreen. 2005. *For Space.* London: Sage Publications Ltd.

Mastai, Judith. 2007. 'There Is No Such Thing as a Visitor'. In *Museums after Modernism: Strategies for Engagement*, edited by Griselda Pollock and Joyce Zemans, 173–177. Oxford: Blackwell Publishing.

MDR. 2020. 'Sächsischer Gedenkstättenchef Reiprich ab sofort suspendiert'. www.mdr.de/nachrichten/sachsen/gedenkstaetten-chef-reiprich-suspendiert-100.html.

Mitchell, Donald. 2000. *Cultural Geography: A Critical Introduction.* Malden, MA: John Wiley & Sons.

Möller, Lena. 2019. *'Auf Stätten des Leids Heime des Glücks': Die Siedlung am Vogelherd auf dem Areal des ehemaligen KZ Flossenbürg und ihre Emotionalisierung als Wohn- und Gedächtnisort.* Münster: Waxmann Verlag.

Morsch, Günter. 2018. 'Sachsenhausen Concentration Camp: Anniversary of Liberation. 22. 04.2018'. www.dw.com/en/sachsenhausen-concentration-camp-anniversary-of-liberation/a-43483448.

Neiman, Susan. 2019. *Learning from the Germans. Confronting Race and the Memory of Evil.* London: Allen Lane.

Nowthisnews. 2018. 'President Trump Weirdly Praised the "Beautiful Sight" of Barbed Wire as U.S. Soldiers Installed It across the Southern Border'. 4 December. https://twitter.com/nowthisnews/status/1059223585166774273.

Oswalt, Philipp, and Stephanie Oswalt. 2000. 'Entwurf zur Gestaltung der erweiterten Gedenkstätte Ravensbrück'. In *Das Mädchenkonzentrationslager Uckermark*, edited by Katja Limbächer, Maike Merten, and Bettina Pfefferle, 280–292. Münster: Unrast Verlag.

Perry, Deborah L. 2012. *What Makes Learning Fun?: Principles for the Design of Intrinsically Motivating Museum Exhibits.* Plymouth: AltaMira Press.

Popescu, Diana I. 2016. 'Post-Witnessing the Concentration Camps: Paul Auster's and Angela Morgan Cutler's Investigative and Imaginative Encounters with Sites of Mass Murder'. *Holocaust Studies* 22 (2–3): 274–288.

Rapson, Jessica. 2015. *Topographies of Suffering: Buchenwald, Babi Yar, Lidice.* New York: Berghahn Books.

Reynolds, Daniel. 2018. *Postcards from Auschwitz-Holocaust Tourism and the Meaning of Remembrance.* New York: New York University Press.

Rothberg, Michael. 2020. 'Das Gespenst des Vergleichs'. Translated by Kathrin Hadeler. www.goethe.de/prj/lat/de/dis/21864662.html.

Ryan, Richard M., and Edward L. Deci. 2000. 'Intrinsic and Extrinsic Motivations: Classic Definitions and New Directions'. *Contemporary Educational Psychology* 25 (1): 54–67.

Rydén, Johanna Bergqvist. 2018. 'When Bereaved of Everything: Objects from the Concentration Camp of Ravensbrück as Expressions of Resistance, Memory, and Identity'. *International Journal of Historical Archaeology* 22 (3): 511–530. https://doi.org/10.1007/s10761-017-0433-2.

Sandahl, Jette. 2020. 'In Place of Denial: The Skills of Honoring Difference and Disagreement'. In *Das Museum der Zukunft. 43 neue Beiträge zur Diskussion über die Zukunft des Museums*, edited by Schnittpunkt and Joachim Baur, 235–240. Bielefeld: Transcript Verlag.

Sawicki, Pawel. 2018. 'Research Centre for Economics at Memorial Sites'. *Memoria* 5: 24–27.

Schult, Tanja. 2020. 'Christoph Mayer's the Invisible Camp: Audio Walk Gusen'. *Liminalities: A Journal of Performance Studies* 16 (1). http://liminalities.net/16-1/loss.html.

Sebald, Gerd. 2011. 'Einleitung: Zur Selektivität von sozialen Erinnerungen'. In *Soziale Gedächtnisse. Selektivitäten in Erinnerungen an die Zeit des Nationalsozialismus*, edited by Gerd Sebald, René Lehmann, Monika Malinowska, Florian Öchsner, Christian Brunnert, and Johanna Frohnhöfer, 9–22. Bielefeld: Transcript Verlag.

Siebeck, Cornelia. 2014. '"The Universal Is an Empty Place": Nachdenken über die (Un) Möglichkeit demokratischer KZ-Gedenkstätten'. In *Ereignis & Gedächtnis – Neue Perspektiven auf die Geschichte der nationalsozialistischen Konzentrationslager*, edited by Imke Hansen, Enrico Heitzer, and Katarzyna Nowak, 217–253. Berlin: Metropol Verlag.

Simon, Nina. 2010. *The Participatory Museum.* Santa Cruz, CA: Museum 2.0.

Sommer, Robert. 2020. 'Perverse Sadisten in SS-Uniform. Zum sexualisierten Bild von TäterInnen in Nazi-Pulp und -Pornografie'. In *Ravensbrück denken: Gedenk- und Erinnerungskultur im Spannungsfeld von Gegenwart und Zukunft. Festschrift zum Abschied von Insa Eschebach als Leiterin der Mahn- und Gedenkstätte Ravensbrück*, edited by Sabine Arend and Petra Fank, 114–122. Berlin: Metropol Verlag.

Sontag, Susan. 2013. *Regarding the Pain of Others.* London: Penguin Books.

Sumartojo, Shanti. 2019. 'Sensory Impact: Memory, Affect and Sensory Ethnography at Official Memory Sites'. In *Doing Memory Research: New Methods and Approaches,* edited by Danielle Drozdzewski and Caroline Birdsall, 21–37. Singapore: Palgrave Macmillan.

———. 2020. 'Lieux de Mémoire through the Senses: Memory, State-Sponsored History, and Sensory Experience'. In *The Routledge Handbook of Memory and Place*, edited by Sarah De Nardi, Hilary Orange, Steven High, and Eerika Koskinen-Koivisto, 249–253. London; New York: Routledge.

Thöne, Eva. 2018. 'Aufarbeitung der NS-Zeit: "Eine Verhöhnung der Opfer"'. *Spiegel Online*, 11 June 2018. www.spiegel.de/kultur/gesellschaft/aufarbeitung-der-ns-zeiteine-verhoehnung-der-opfer-a-1211778.html.

USC Institute for Creative Technologies. 2017. 'Keeping Holocaust Survivor Testimonies Alive: Through Holograms'. Accessed 29 March 2021. https://ict.usc.edu/news/keeping-holocaust-survivor-testimonies-alive-through-holograms/.

Van der Laarse, Rob. 2013. 'Beyond Auschwitz? Europe's Terrorscapes in the Age of Postmemory'. In *Memory and Postwar Memorials: Confronting the Violence of the Past*, edited by Marc Silberman and Florence Vatan, 71–92. New York: Palgrave Macmillan.

Wagner, Jens-Christian. 2017. 'Gedenkstättenarbeit in Deutschland seit 1945: Eine Erfolgsgeschichte?'. *Netzwerk Erinnerung und Zukunft Hannover e.V.* (September): 5–10.

———. 2018. 'Simulierte Authentizität? Chancen und Risiken von Augemted and Virtual Reality an Gedenkstätten'. *Gedenkstättenrundbrief* 196: 3–9.

Watts, Jonathan. 2018. 'Red List Research Finds 26,000 Global Species under Extinction Threat'. *The Guardian*, 5 July 2018,. www.theguardian.com/environment/2018/jul/05/red-list-research-finds-26000-species-under-extinction-threat.

Whigham, Kerry. 2020. 'Reading the Traces: Embodied Engagement with the Past at Three Former Nazi Concentration Camps'. *Holocaust Studies* 26 (2): 221–240.

Zalewska, Maria. 2018. 'Selfies from Auschwitz: Rethinking the Relationship Between Spaces of Memory and Places of Commemoration in the Digital Age'. *Studies in Russian, Eurasian and Central European New Media* (blog), 13 June. www.digitalicons.org/issue18/selfies-from-auschwitz-rethinking-the-relationship/.

Zelizer, Barbie. 1998. *Remembering to Forget: Holocaust Memory through the Camera's Eye*. Chicago: University of Chicago Press.

Index

For Product Safety Concerns and Information please contact our EU
representative GPSR@taylorandfrancis.com
Taylor & Francis Verlag GmbH, Kaufingerstraße 24, 80331 München, Germany